INTERNATIONAL INTERIOR DESIGN YEARBOOK 2016

2016
国际室内设计年鉴 5

HOTEL/CLUB COSMETOLOGY
酒店/美容 会所

本书编委会 编

中国林业出版社
China Forestry Publishing House

图书在版编目（C I P）数据

国际室内设计年鉴. 2016. 酒店、美容、会所 / 《国际室内设计年鉴》编委会编. -- 北京 : 中国林业出版社,
2016.4

ISBN 978-7-5038-8449-8

Ⅰ.①国… Ⅱ.①国… Ⅲ.①商业建筑－室内装饰设计－世界－2016－年鉴 Ⅳ.①TU238-54

中国版本图书馆CIP数据核字(2016)第057469号

本书编委会

◎ 编委会成员名单

丛书主编：柳素荣

编写成员：陈向明　陈治强　董世雄　冯振勇　朱统菁

◎ 丛书策划：北京和易空间文化传播有限公司

◎ 特别鸣谢：《室内设计与装修》杂志社

◎ 装帧设计：北京睿宸弘文文化传播有限公司+LOMO红绿

中国林业出版社 · 建筑家居出版分社

责任编辑：王思源 纪 亮

出版：中国林业出版社 （100009 北京西城区德内大街刘海胡同 7 号）
网址：lycb.forestry.gov.cn
电话：（010）8314 3518
发行：中国林业出版社
印刷：北京利丰雅高长城印刷有限公司
版次：2017年2月第1版
印次：2017年2月第1次
开本：230mm × 305mm　1/16
印张：13.5
字数：200千字
定价：220.00 元

CONTENTS 目录

INTRO-
DUCTION

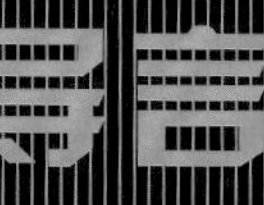

/ RECORD THE EXCELLENCE PUBLISH THE QUINTESSNCE

/ 记录精英 传播经典

张先慧 /Zhang Xianhui

中国麦迪逊文化传播机构董事长
中国（广州、上海、北京）广告书店董事长
《麦迪逊丛书》主编
Chairman of China Madison Culture Communication Institutions
President of China(Guangzhou,Shanghai,Beijing) Advertising Bookshop
Chief Editor of "The Madison Series"

人的一生，绝大部分时间是在室内度过的。因此，人们设计创造的室内环境，必然会直接关系到人们室内生活、生产活动的质量，关系到人们的安全、健康、效率、舒适，等等。随着人们生活水平和审美能力的不断提高，人们更加注重生活环境的设计，对于室内设计的要求更加严格，需求也日益多样化、个性化。这就要求设计师一定要牢牢把握住时代的脉搏和潮流，以独特的眼光，运用与众不同的角度和表现手法进行创意性的设计，以满足人们对室内设计的需求。

然而，一件好的设计作品，不仅与设计师的专业素质和文化艺术素养等联系在一起，更离不开对他人成功经验的借鉴，为此，《国际室内设计年鉴2011》应运而生。

本年鉴秉持以中国大陆、中国香港、中国台湾为主，兼容其他国家与地区参与的原则，主张以创新与发展作为室内设计创作的主旋律，以科学与艺术相结合的审美眼光审视室内设计作品，力求打造全球最具影响力的室内设计行业年鉴，并使其成为各国设计师可以借鉴的经典书籍。

本年鉴征稿消息发出后，世界各地的设计机构与设计师都踊跃参与，大量投稿，投稿数量之多完全出乎我们的意料，最终本年鉴以一套五册的形式面世。

我们用年鉴的形式把当代最具价值的室内设计作品记录下来，传播开去，意在对室内设计文化予以保存的同时，也为读者提供了解当代设计状况及思想交流的平台。

“记录精英，传播经典”，这是《麦迪逊丛书》的宗旨。

希望业界朋友继续关注与支持我们。

One's lifetime mostly passes through in the interior. Therefore, the interior environment will directly involve quality of people's interior life, activities, people's safety, health, efficiency, comfort and so on. Along with the continuous improvement of people's living standard and aesthetic capacity, people pay more attention to living environment design, and their requirement for interior design is more strict, increasingly diverse and personalized. This requires that the designer firmly grasp the pulse of the times and trends, with the special insight, to use the different angles and methods of performance for creative design, in order to meet the needs of people's interior design requirement.

A masterpiece requires not only the link of the designer's professional quality and cultural art accomplishment, but also others' successful experiences. For this reason, " International Interior Design Yearbook 2010" is born at this right moment.

This yearbook gives priority to China Mainland, China Hong Kong and China Taiwan and pays much attention to other countries and areas, and it upholds the spirit that innovation and development should be the theme of interior design and that interior design works should be evaluated in a scientific and artistic perspective. Aiming at becoming the most influential global yearbook of interior design, this book is a classical one in the eyes of designers all over the world.

After the announcement of draft-collecting was spread, we have received so many contributions from the designers and organizations of almost every country. The number was so surprising. Finally, the yearbook is published in a set of five books.

We present the most valuable contemporary interior designs through publishing this yearbook in order to preserve the interior designing culture and provide a platform for readers to know about contemporary designing improvements and to communicate with each other.

"Record Excellence Works, Spread Classical Works" is the tenet of "Madison Series".

It will be our privilege to have your appreciation and support.

HOTEL
酒店

瑞丽时尚酒店

RUILI FASHION HOTEL

项目资料：
设计单位：创所未想商业空间一体化设计
设计总监（主创）：余涛
参与设计师：王先岭 杨殿玺 沈燕
项目地址：苏州市十全街相王路
主要材料：装饰板、PVC装饰管、黑镜 仿真草皮、马赛克

Project Information:
Design Unit: Chuangsuoweixiang Cemmercial Space Union Design
Design Director (Main Director): Yu Tao
Involed Designer: Wang Xianling, Yang Dianxi, Shen Yan
Project Address: Suzhou Shiquan Street Xiangwang Road
Materials: decorative plate, PVC pipe, Black mirror, Simulation turf, mosaic

To pot ism

HILTON SAN DIEGO BAYFRONT酒店

HILTON SAN DIEGO BAYFRONT HOTEL

项目资料：
设计单位：John Portman & Associates
设计建筑：John Portman & Associates for the exterior and public areas of the hotel
建筑助理：Joseph Wong Design Associates for the guestrooms in the hotel
摄影师：Jim Brady, Courtesy of Hilton San Diego Bayfront, Courtesy of John Portman & Associates
开发人员：Phelps Portman San Diego, LLC
总承包：Hensel Phelps Construction Company
客户：One Park Boulevard, LLC
项目地址：San Diego, California USA
景观设计：Sasaki Associates, Inc.
公共艺术：Ned Khan, Norie Sato, Nance O'Banion and John Portman
主要材料：釉面幕墙与金属板、混凝土等.
完成时间：2009年3月

Project information:
Design Unit: John Portman & Associates
Design Architect: John Portman & Associates for the exterior and public areas of the hotel
Associate Architect: Joseph Wong Design Associates for the guestrooms in the hotel
Photographer: Jim Brady, Courtesy of Hilton San Diego Bayfront, Courtesy of John Portman & Associates
Developers: Phelps Portman San Diego, LLC
General Contractor: Hensel Phelps Construction Company
Client: One Park Boulevard, LLC
Address: San Diego, California USA
Landscape Architect: Sasaki Associates, Inc.
Featured Public Artists: Ned Khan, Norie Sato, Nance O'Banion and John Portman
Materials: Glazed Curtainwall with Metal Panels, Concrete, etc.
Completion: March, 2009

Hilton

墨西哥MAROMA海滩度假酒店

SECRETS MAROMA BEACH RIVIERA CANCUN

项目资料：
设计单位：HKS Inc.
摄影师：Secrets Maroma Resort

Project information:
Design Unit: HKS Inc.
Photographer: Secrets Maroma Resort

绍兴咸亨新天地大酒店

SHAOXING XIANHENG NEW EARTH HOTEL

项目资料：

设计单位：杭州陈涛室内设计有限公司

设计师：陈涛 陈旭如 丁永钞 黄珏 王仁洪

Project Information:

Design Unit: Hangzhou Chentao Interior Design Company

Designer: Chen Tao,Chen Xuru, Ding Yongchao ,Wang Renhong

群仙會

华都酒店

WALDO HOTEL

项目资料：

设计单位：德坚设计
摄影师：Mr. Kinney Chan
项目地址：澳门新口岸友谊大马路
面积：215000m²

Project information:

Design Unit: Kinney Chan and Associates
Photographer: Mr. Kinney Chan
Project Address: Avenida de Amizade, Macau
Area: 215000sqm

CASHIER

杭州西湖 温德姆豪庭大酒店

WYNDHAM GRAND PLAZA ROYALE WEST LAKE HANGZHOU

项目资料：
设计单位：J2-STUDIO/厚华顾问设计有限公司
设计师：冯厚华
项目地址：浙江省杭州市西湖边
面积：45000m²
主要材料：磨沙不锈钢、金箔、大理石
完成时间：2009年11月

Project information:
Design Unit: J2-STUDIO/ Hou Hua Consultant Design Co,.Ltd
Designer: Map.Feng
Project Address: Hangzhou City, Zhejiang Province
Area: 45000sqm
Materials: Brushed stainless steel, gold, marble
Completion : November, 2009

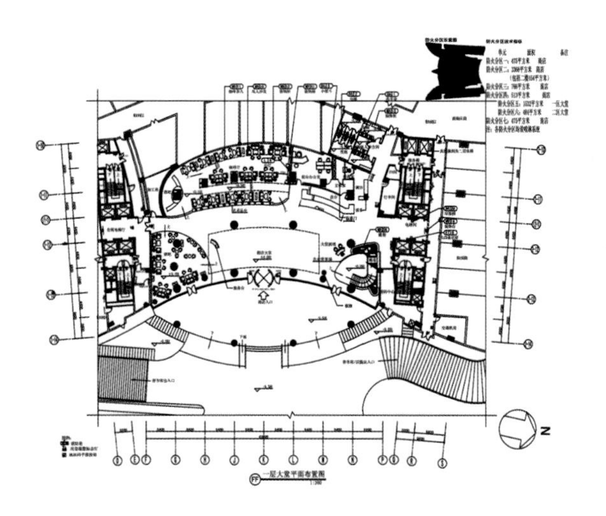

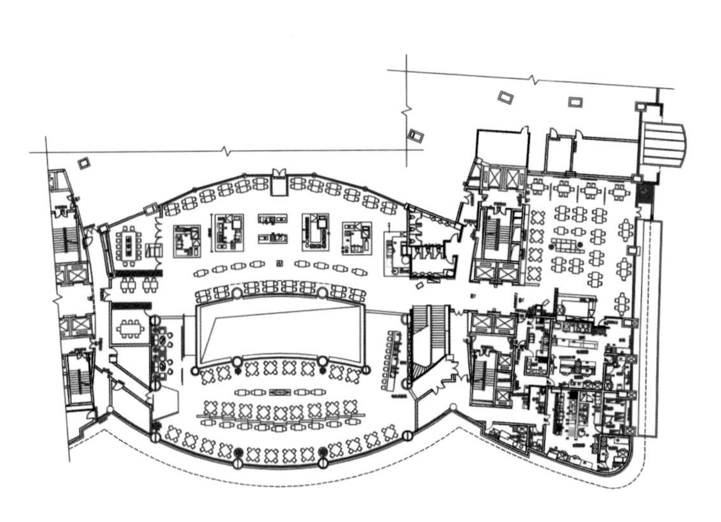

二层自助、风味餐厅平面布置变更图
1:200

HOTEL CALDOR

HOTEL CALDOR

项目资料：
设计单位：SÕHNE & PARTNER Architects
设计团队：Thomas Bärtl, Michael Prodinger , Guido Trampitsch
摄影师：Severin Wurnig
客户：Martin Reichard

Project information:
Design Unit: SÕHNE & PARTNER Architects
Design team: Thomas Bärtl, Michael Prodinger , Guido Trampitsch
Photographer: Severin Wurnig
Client: Martin Reichard

VENETIAN MACAO RESORT HOTEL

VENETIAN MACAO
RESORT HOTEL

项目资料：
设计单位：HKS Inc.
摄影师：Venetian Macao-Resort-Hotel

Project information：
Design Unit：HKS Inc.
Photographer：Venetian Macao-Resort-Hotel

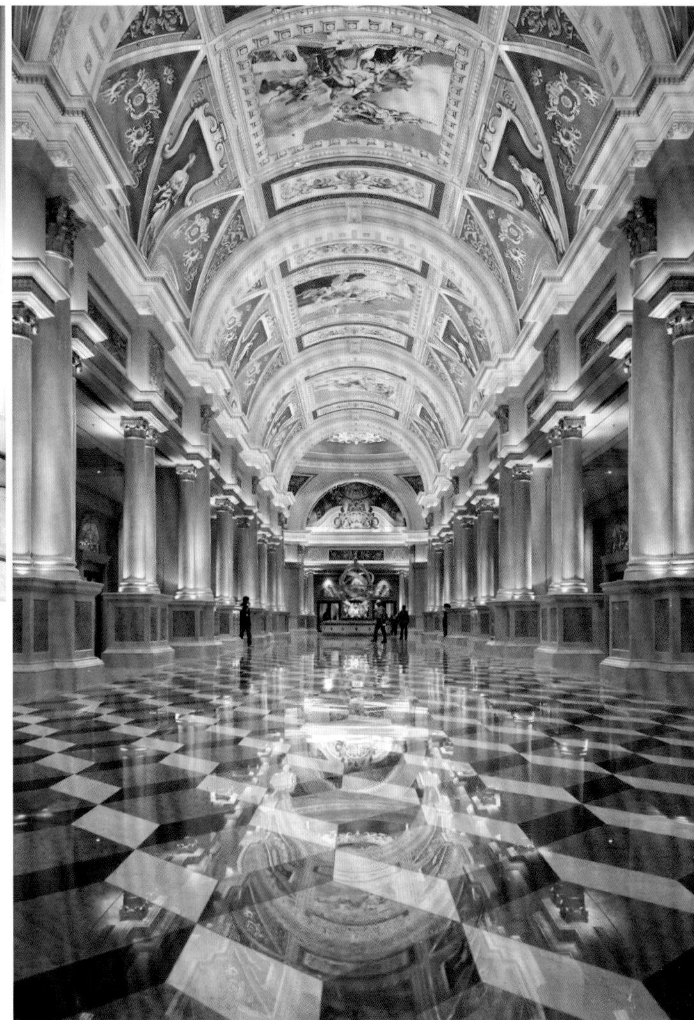

北京首都国际机场希尔顿酒店

BEIJING CAPITAL INTERNATIONAL AIRPORT HILTON HOTEL

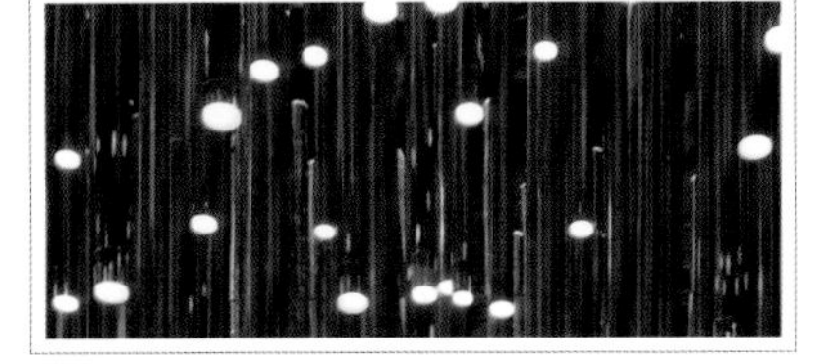

项目资料：
设计单位：深圳市黑龙室内设计有限公司
设计总监（主创）：王黑龙
参与设计团队：黑龙设计
摄影师：李伟
主要材料：索菲亚米黄、伊朗白洞、麦哥利紫影、玫瑰梨木、仿古铜花格
地址：北京首都国际机场T3航站楼
面积：2.58万m^2
完成时间：2009年12月

Project information:
Design Unit: Shenzhen Black Dragon Interior Design Co., Ltd.
Design Director: Wang Heilong
Involved Design Team: Black Dragon Design
Photographer: Li Wei
Materials: Sofia Beige, Iran white hole, Maige Li purple shadow, Rose Lei, antique copper lattice
Project Address: Beijing Capital International Airport T3 Terminal
Area: 2.58 Million sqm
Completion: December,2009

常州香树湾花园酒店一期

CHANGZHOU XIANGSHU BAY GARDEN HOTEL STAGE ONE

项目资料：

设计单位：深圳市黑龙室内设计有限公司
设计总监（主创）：王黑龙
参与设计师：黑龙设计
摄影师：文宗博
项目地点：江苏常州新区汉江路2号
面积：1.56 万m^2
主要材料：砂岩、花岗石、板岩、马赛克、防腐木、酸枝木、紫檀木、斑马木、艺术涂料
完成时间：2010年1月

Project Information:

Design Unit: ShenZhen Black Dragon Iinterior Design Co.Ltd.
Design Director (Main Director): Wang Heilong
Involed Designer Team: Black Dragon Design
Photographer: Wen Zongbo
Project Address: Jiangsu ChangZhou New District Hanjiang Road No.2
Area: 1.56Millionsqm
Materials: Sandstone,granite,slate,mosaic,wood preservative,rosewood,zebra wood,artpaint
Completion: January,2010

CAPELLA PEDREGAL RESORT

CAPELLA PEDREGAL RESORT

项目资料：
设计单位：HKS Inc.
摄影师：Robert Reck

Project information:
Design Unit: HKS Inc.
Photographer: Robert Reck

AFRICA
Michael Poliza

EAST HONG KONG

EAST HONG KONG

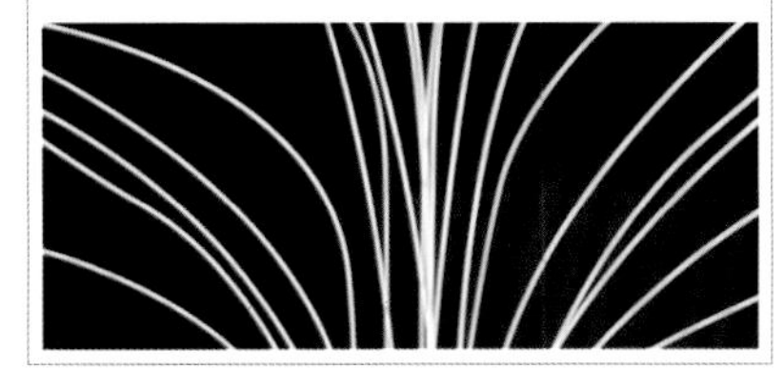

项目资料：
设计单位：CI3 Architects Limited.
客户：Swire Properties
项目地址：香港
面积：17,000m²
完成时间：2010年

Project information：
Design Unit：CI3 Architects Limited.
Client：Swire Properties
Address：Hong Kong
Area：17,000 m
Completion：2010

east
HONG KONG

二十九

河北宽城酒店

HEBEI KUANCHENG HOTEL

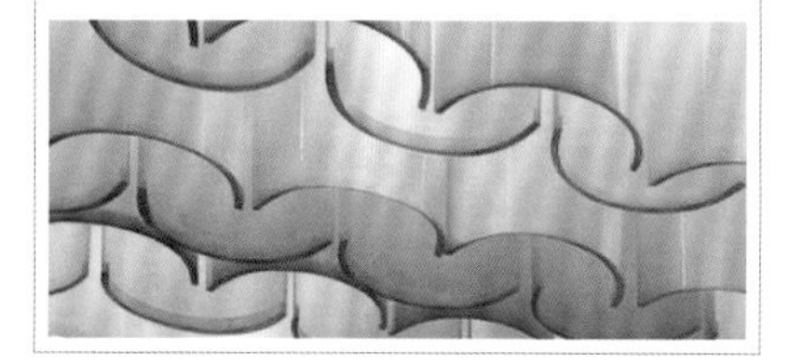

项目资料：

设计单位：青岛易锋建筑装饰设计有限公司
设计总监（主创）：王立锋
参与设计团队：李涛 谭政
摄影师：文宗博 潘宇峰
项目地址：河北宽城
主要材料：石材、地砖

Project information:

Design Unit: Qingdao Yifeng Architectural Decoration Design Co., Ltd.
Design Director(Main Director): Wang Lifeng
Involed Design Team: Li Tao, Tan Zheng
Photographer: Wen Zongbo, Pan Yufeng
Project Address: Hebei Kuancheng
Materials: Stone, Tiles

徐州小南湖凯莱度假酒店

XUZHOU XIAONAN LAKE KAILAI HOLIDAY HOTEL

项目资料：
设计单位：北京丽贝亚建筑装饰工程有限公司
设计总监（主创）：贺光宇 刘珠鹏
参与设计团队：金哲秀 屈勇 贾海燕 迟凯 曹姗娜 卢迪 张祎 杜春亮 马青
摄影师：徐盟 李庚生
项目地址：江苏省徐州市小南湖
主要材料：珍珠木、柚木、梵蒂米黄石材、新帝王石材、浅啡网石材、阿波罗石材、法国流金石材、金箔、银箔、席纹壁纸、艺术壁纸、铁板皂化、铜板、羊皮

Project information:
Design Unit: Beijing Libeiya Architectural Decoration Engineering Co., Ltd.
Design Director(Main Director): He Guangyu, Liu Zhupeng
Involved Design Team: Jin Zhexiu, Qu Yong, Jia Haiyan, Chi Kai, Cao Shanna, Lu Di, Zhang Hui, Du Chunliang, Ma Qing
Photographer: Xu Meng, Li Gengsheng
Project Address: Jiangsu Xuzhou Small South Lake
Materials: Pearl wood, teak, the Vatican Timmy yellow stone, Xindiwang stone, light brown network of stone, Apollo stone, the French golden stone, gold foil, silver foil, Xi pattern wallpaper, art wallpaper, iron, copper, sheepskin

PORTO PALÁCIO HOTEL

PORTO PÁLACIO HOTEL

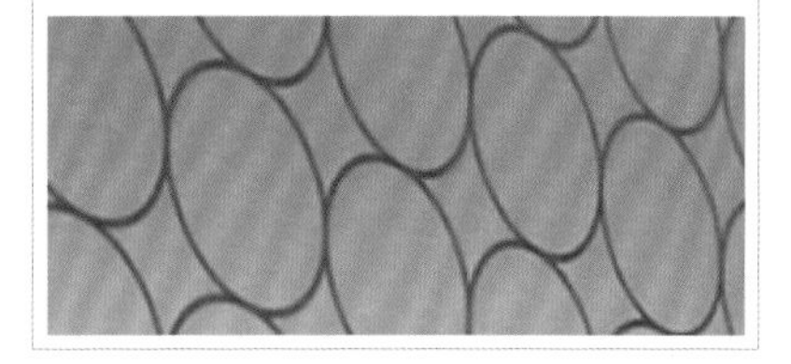

项目资料：

设计单位：Plajer & Franz Studio
项目管理：michael bertram
摄影师：ken schluchtmann
客户：sonae tourismo

Project information:

Design Unit: Plajer & Franz Studio
Project manager: Michael bertram
Photographer: ken schluchtmann
Client: sonae tourismo

Porto Palácio
HOTEL

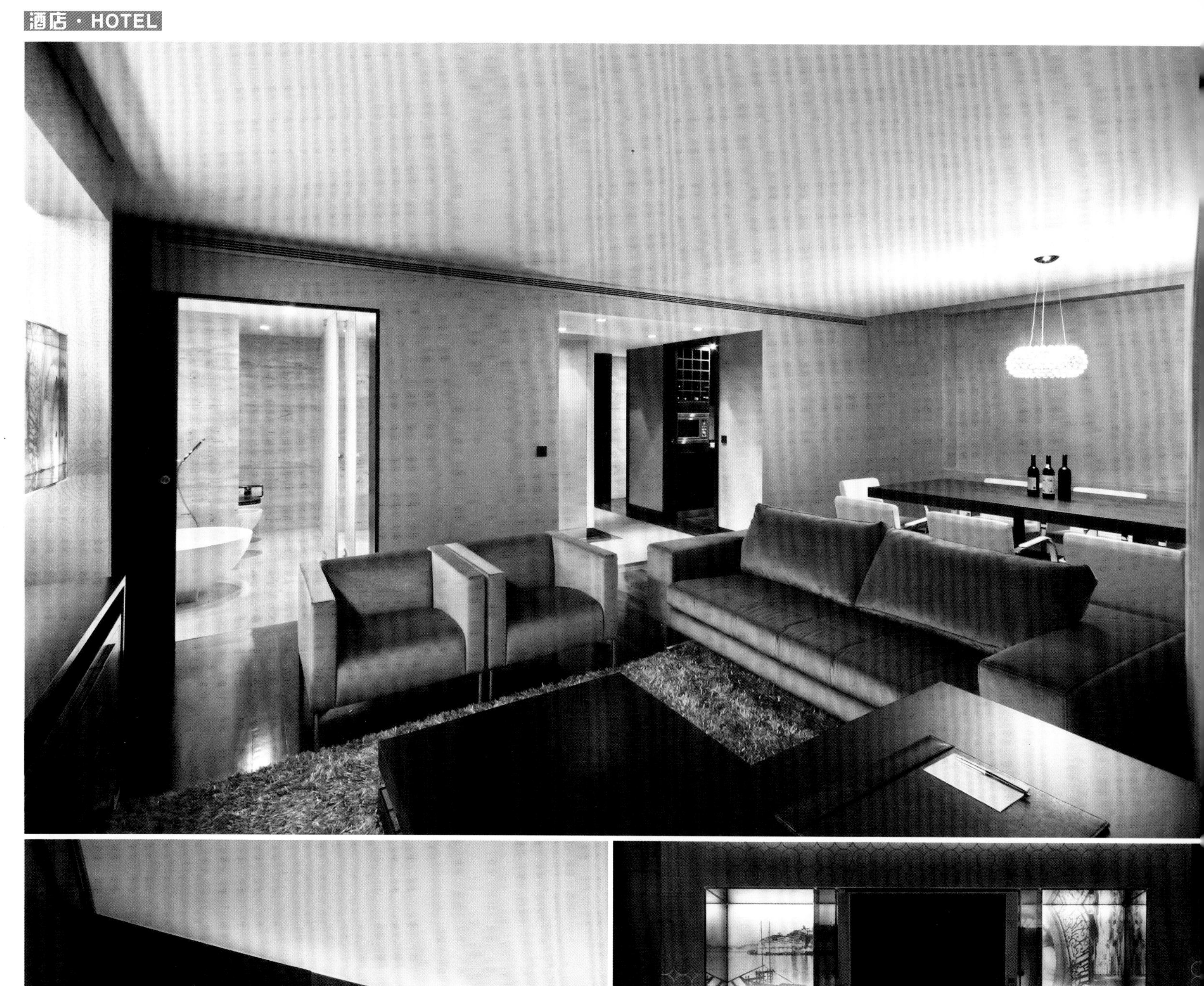

EOS ACAPULCO

EOS ACAPULCO

项目资料：

设计单位：DPGa
设计师：ARCH. DANIEL PEREZ GIL DIR.
摄影师：Hector Armando Herrera
客户：Promotora tres palos S de RL de CV
面积：12000m²

Project information:

Design Unit: DPGa
Designer: ARCH. DANIEL PEREZ GIL DIR.
Photographer: Hector Armando Herrera
Client: Promotora tres palos S de RL de CV
Area: 12,000sqm

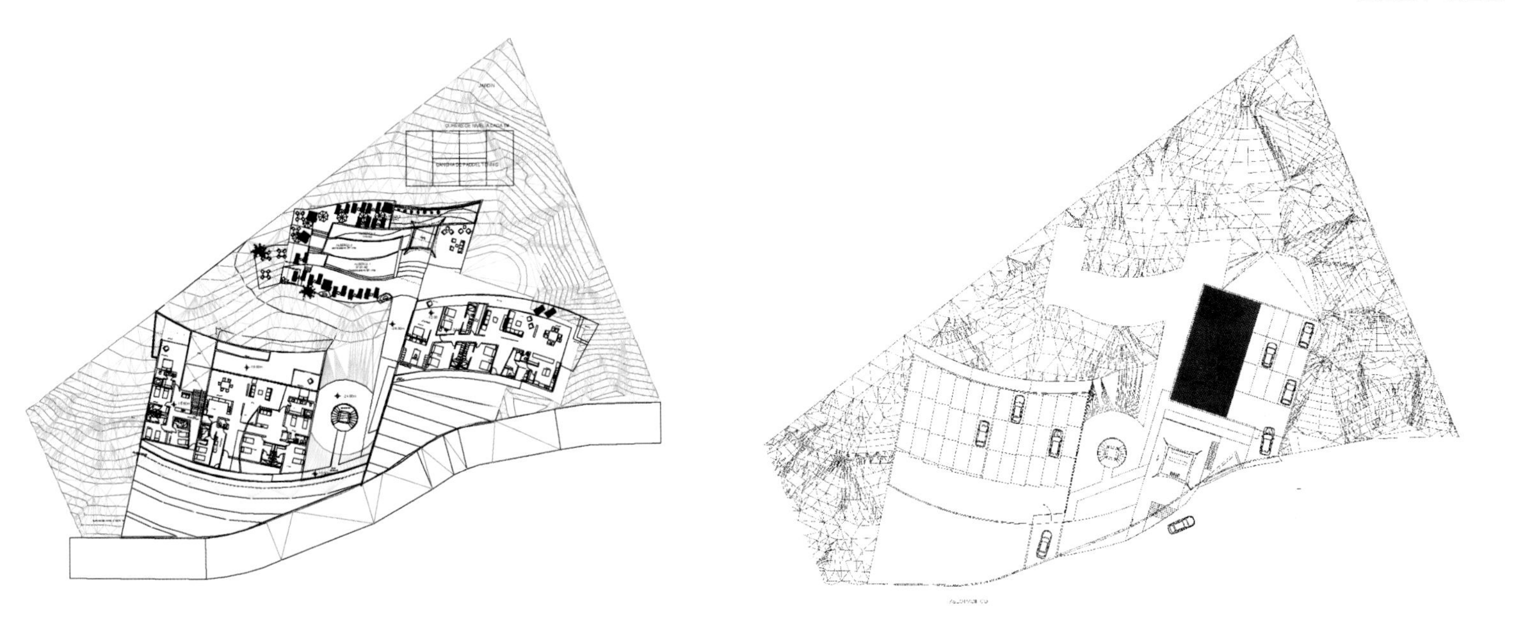

台北沐兰
精品旅馆(203房)

TAIPEI MULAN
BONTIQUE HOTEL

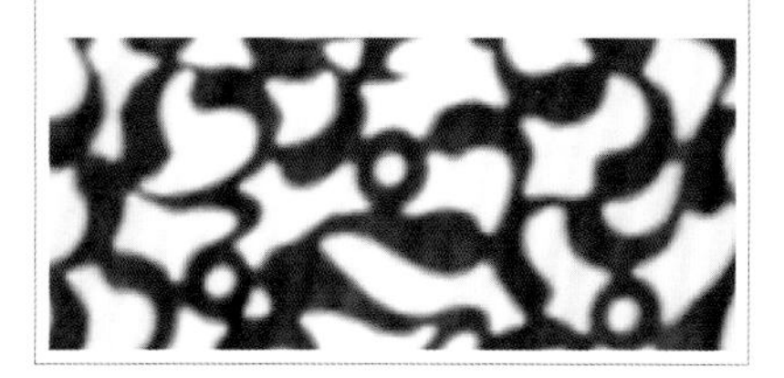

项目资料：
设计单位：杨焕生建筑室内设计事务所
设计总监：杨焕生
参与设计团队：王慧静 郭士豪
摄影师：刘俊杰
项目地址：台北市

Project information:
Design Unit: Yang Huansheng
Architectural Interior Design Office
Design Director: Yang Huansheng
Involed Design Team: Wang Huijing, Guo Shihao
Photographer: Liu Junjie
Project Address: Taipei

台北沐兰
精品旅馆(505房)

TAIPEI MULAN BONTIQUE HOTEL

项目资料：

设计单位：杨焕生建筑室内设计事务所
设计总监：杨焕生
参与设计团队：王慧静 郭士豪
摄影师：刘俊杰
项目地址：台北市
面积：100 m^2
主要材料：石材、橡木、仿古漆、进口马赛克、茶镜

Project information:

Design Unit: Yang Huansheng Architectural Interior Design Office
Design Director: Yang Huansheng
Involed Design Team: Wang Huijing, Guo Shihao
Photographer: Liu Junjie
Project Address: Taipei
Area: 100 sqm
Materials: Stone, oak, antique paint, imported mosaic, tea mirror

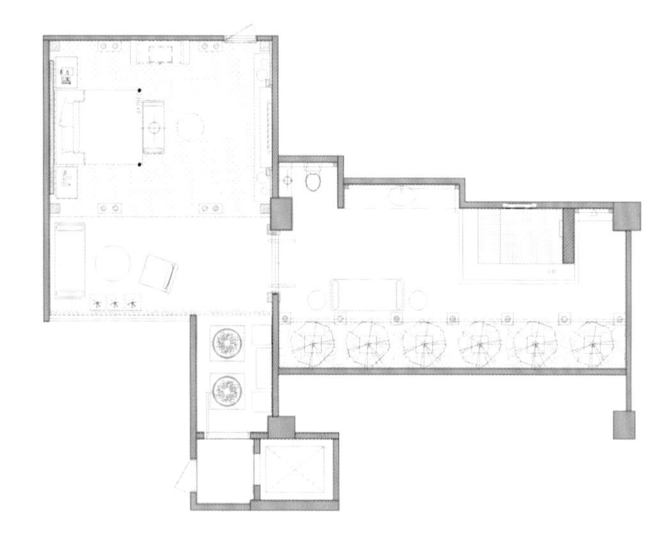

台北沐兰精品旅馆（509房）

TAIPEI MULAN BONTIQUE HOTEL

项目资料：

设计单位：杨焕生建筑室内设计事务所
设计总监：杨焕生
参与设计团队：王慧静 郭士豪
摄影师：刘俊杰
面积：130m²
主要材料：凿面石材、黑檀木、裱布、玻璃、贝壳马赛克
项目地址：台北市

Project information:

Design Unit: Yang Huansheng Architectural Interior Design Office
Design Director: Yang Huansheng
Involed Design Team: Wang Huijing, Guo Shihao
Photographer: Liu Junjie
Area: 130sqm
Materials: Cutting surface stone, ebony, mounted cloth, glass, shell mosaic
Project Address: Taipei

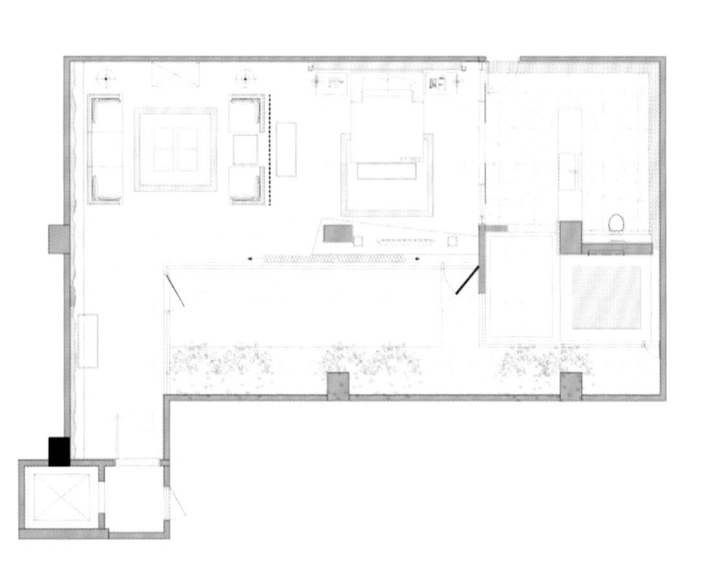

台北沐兰 精品旅馆(602房)

TAIPEI MULAN BONTIQUE HOTEL

项目资料：

设计单位：杨焕生建筑室内设计事务所
设计总监：杨焕生
参与设计团队：王慧静 郭士豪
摄影师：刘俊杰
项目地址：台北市

Project information:

Design Unit: Yang Huansheng Architectural Interior Design Office
Design Director: Yang Huansheng
Involed Design Team: Wang Huijing, Guo Shihao
Photographer: Liu Junjie
Project Address: Taipei

台北沐兰精品旅馆(606)

TAIPEI MULAN
BONTIQUE HOTEL

项目资料：

设计单位：杨焕生建筑室内设计事务所
设计总监：杨焕生
参与设计团队：王慧静 郭士豪
摄影师：刘俊杰
项目地址：台北市

Projet information:

Design Unit: Yang Huansheng Architectural Interior Design Office
Design Director: Yang Huansheng
Involed Design Team: Wang Huijing, Guo Shihao
Photographer: Liu Junjie
Project Address: Taipei

台北淋兰精品旅馆（609）

TAIPEI MULAN BONTIQUE HOTEL

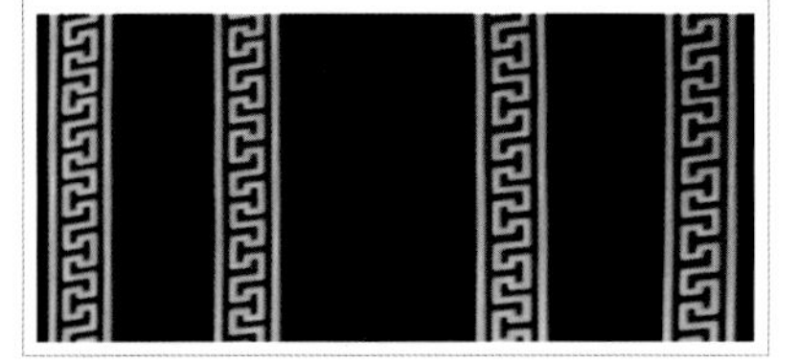

项目资料：

设计单位：杨焕生建筑室内设计事务所
设计总监：杨焕生
参与设计团队：王慧静 郭士豪
摄影师：刘俊杰
项目地址：台北市

Project information:

Design Unit: Yang Huansheng Architectural Interior Design Office
Design Director: Yang Huansheng
Involed Design Team: Wang Huijing, Guo Shihao
Photographer: Liu Junjie
Project Address: Taipei

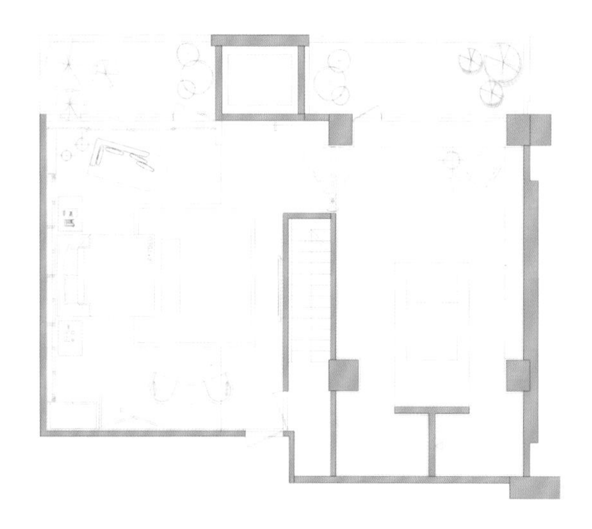

台北沐兰
精品旅馆(610房)

TAIPEI MULAN
BONTIQUE HOTEL

项目资料：
设计单位：杨焕生建筑室内设计事务所
设计总监：杨焕生
参与设计团队：王慧静 郭士豪
摄影师：刘俊杰
项目地址：台北市
面积：120m²
主要材料：石材、白橡木、烤漆、
进口马赛克、茶镜黑境

Project information:
Design Unit: Yang Huansheng
Architectural Interior Design Office
Design Director: Yang Huansheng
Involed Design Team: Wang Huijing, Guo Shihao
Photographer: Liu Junjie
Project Address: Taipei
Area: 120sqm
Materials: Stone, white oak, paint,
imported mosaic, Tea Mirror, black mirror

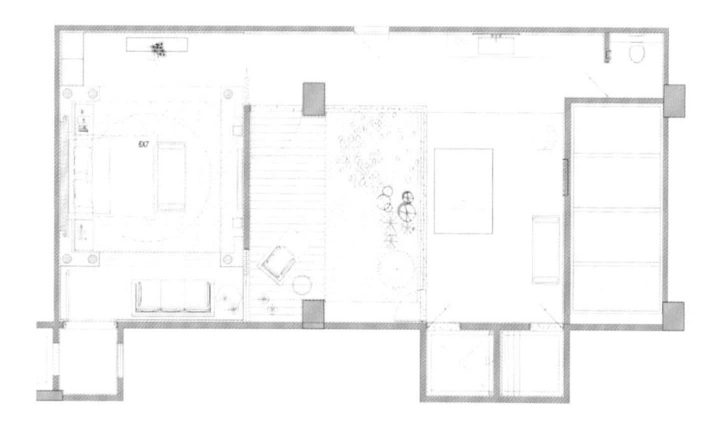

台北沐兰精品旅馆(612房)

TAIPEI MULAN BONTIQUE HOTEL

项目资料：

设计单位：杨焕生建筑室内设计事务所
设计总监：杨焕生
参与设计团队：王慧静 郭士豪
摄影师：刘俊杰
项目地址：台北市
面积：150m²
主要材料：石材、橡木、银箔、马赛克、茶镜

Projet information:

Design Unit: Yang Huansheng Architectural Interior Design Office
Design Director: Yang Huansheng
Involed Design Team: Wang Huijing, Guo Shihao
Photographer: Liu Junjie
Project Address: Taipei
Area: 150sqm
Materials: Stone, Oak, Silver, Masaics, Tea Mirror

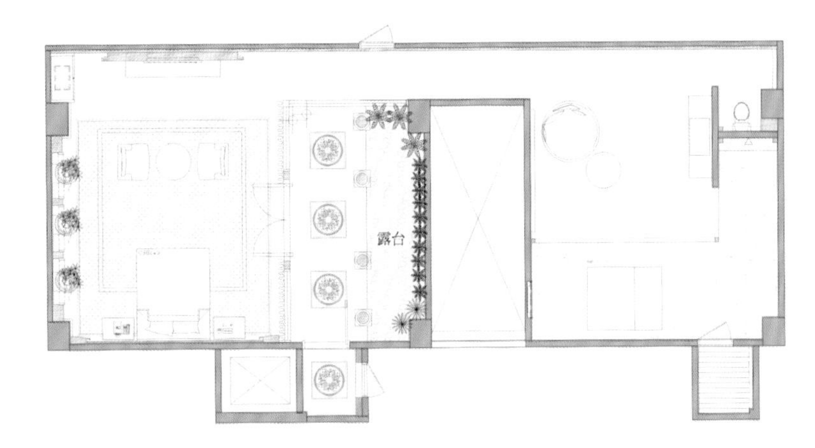

台北沐兰精品旅馆(615)

TAIPEI MULAN BONTIQUE HOTEL

项目资料：

设计单位：杨焕生建筑室内设计事务所
设计总监：杨焕生
参与设计团队：王慧静 郭士豪
摄影师：刘俊杰
项目地址：台北市
面积：150m²
主要材料：石材、橡木、银箔、马赛克、茶镜

Project information:

Design Unit: Yang Huansheng Architectural Interior Design Office
Design Director: Yang Huansheng
Involed Design Team: Wang Huijing, Guo Shihao
Photographer: Liu Junjie
Project Address: Taipei
Area: 150sqm
Materials: Stone, Oak, Silver, Masaics, Tea Mirror

W HOLLYWOOD HOTEL & RESIDENCES

W HOLLYWOOD
HOTEL & RESIDENCES

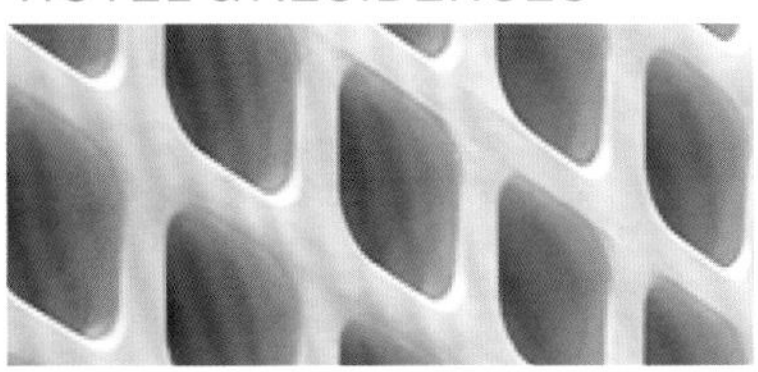

项目资料：
设计单位：HKS, Inc.
摄影师：Blake Marvin, HKS, Inc.

Project information:
Design Unit: HKS, Inc.
Photographer: Blake Marvin, HKS, Inc.

成旅晶赞饭店 · 台北淡水

CHENGLVJINGZAN HOTEL · TAIPEI DANSHUI

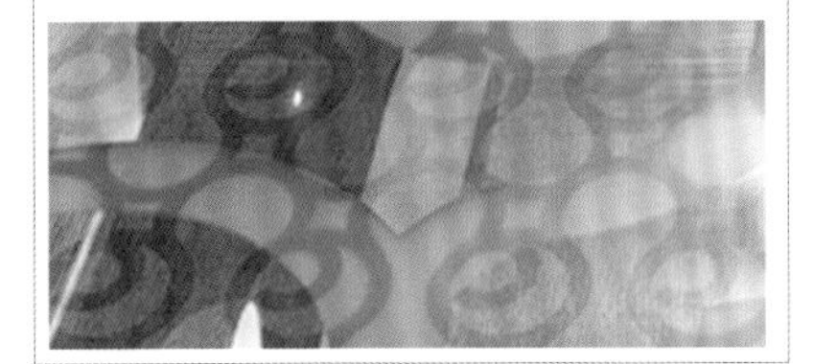

项目资料：
设计单位：颉合室内装修股份有限公司
设计师：江姿莹 黄严仕
参与设计团队：陈冠伃
摄影师：Sam+Yvonne
项目地址：台北县淡水镇中正东路二段91号
主要材料：梧桐木、玻璃、石木

Project information:
Design Unit: Jiahe Interior Decoration Co.,LTD
Designer: Jiang Ziying, Huang Yanshi
Involed Design Team: Chen Guanyu
Photographer: Sam+Yvonne
Project Address: Taipei Danshui Town Zhongzheng East Road Erduan No.91
Materials: Indus wood, glass, stone-wood

UNIVERSUM LOUNGE

UNIVERSUM LOUNGE

项目资料：
设计单位：Plajer & Franz Studio
摄影师：ken schluchtmann
客户：franco francucci
面积：87m²

Project information:
Design Unit: Plajer & Franz Studio
Photographer: ken schluchtmann
Client: Franco francucci
Area: 87sqm

11:01 28

ANDAZ WALL STREET

ANDAZ WALL STREET

项目资料：
设计单位：Rockwell Group
摄影师：Courtesy of Andaz Wall Street
客户：Hyatt Hotels & Resorts,The Hakimian Organization

Project information:
Design Unit: Rockwell Group
Photographer: Courtesy of Andaz Wall Street
Client: Hyatt Hotels & Resorts,The Hakimian Organization

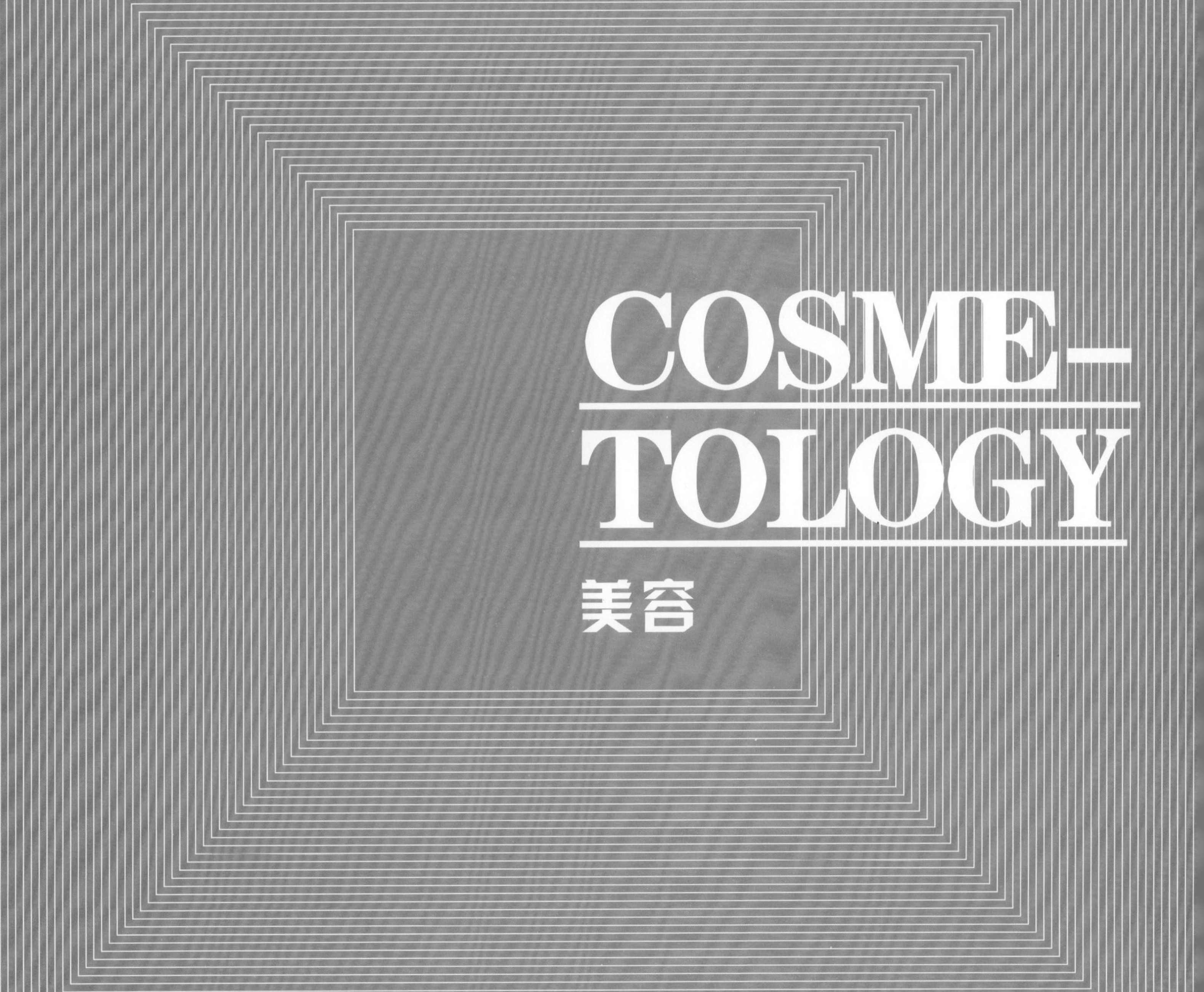
COSME-
TOLOGY
美容

福清素丽娅泰水疗

FUQING QINLIYA THAILAND SPA

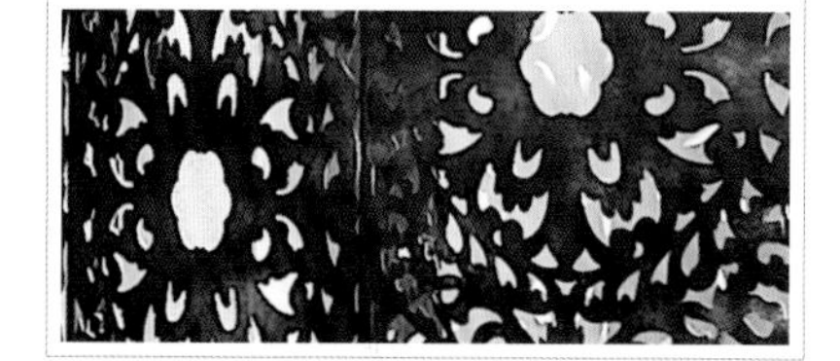

项目资料：

设计单位：福建国广一叶建筑装饰设计工程有限公司
设计师：何华武 李金珍 龚志强 陈亿元
面积：1200m²
主要材料：柚木、金属砖、金碧辉煌、火烧山西黑、手工地毯、硅藻泥、钢片雕花、草编墙纸

Project Information:

Design Unit: Fujian Guoguang Yiye Architectural Decoration Design Engineering Co., Ltd.
Designer: He Huawu, Li Jinzhen, Gong Zhiqiang, Chen Yiyuan
Area: 1200sqm
Materials: Teak, metal tiles, magnificent, fire Shanxi black, handmade carpets, diatom mud, steel carving, straw wallpaper

柏莉雅女子会所

BOLIYAWEMEN CLUB

项目资料：

设计单位：福州林开新室内设计有限公司
设计师：林开新
摄影师：吴永长
项目地址：福州市
面积：400m²
主要材料：软膜、大理石、玻璃、软包
完成时间：2009年3月

Project Information:

Design Unit: Fuzhou Linkaixin Interior Design Co., Ltd.
Designer: Lin Kaixin
Photographer: Wu Yongchang
Project Address: Fuzhou
Area: 400sqm
Materials: Soft membrane, marble, glass, soft pack
Completion: March, 2009

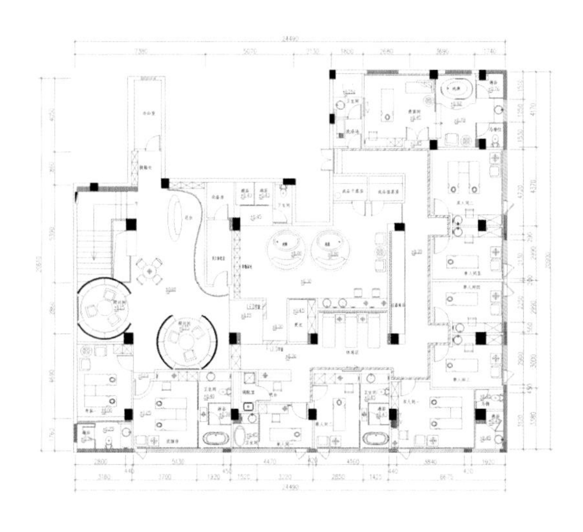

ZZ.SANKT. GEORG

ZZ.SANKT.GEORG

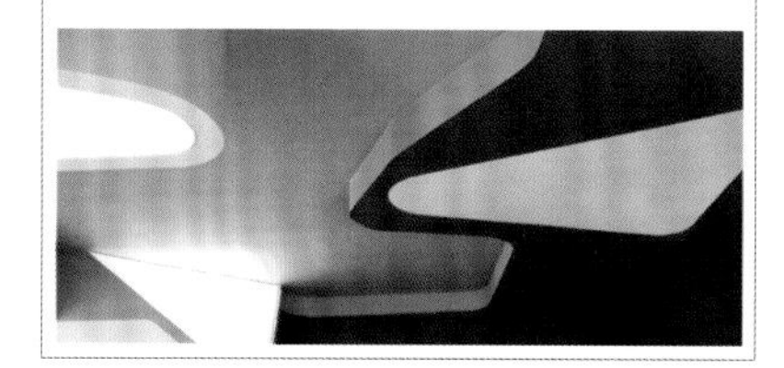

项目资料：

设计单位：J. MAYER H. Architects
设计团队：Jurgen Mayer H., Marcus Blum, Hans Schneider, Wilko Hoffmann
服务工程：Altschul dental GmbH/ Grill&Grill, Frankfurt
摄影师：Ludger Paffrath

Project Information:

Design Unit: J. MAYER H. Architects
Design Team: Jurgen Mayer H., Marcus Blum, Hans Schneider, Wilko Hoffmann
Service Engineers: Altschul dental GmbH/ Grill&Grill, Frankfurt
Photographer: Ludger Paffrath

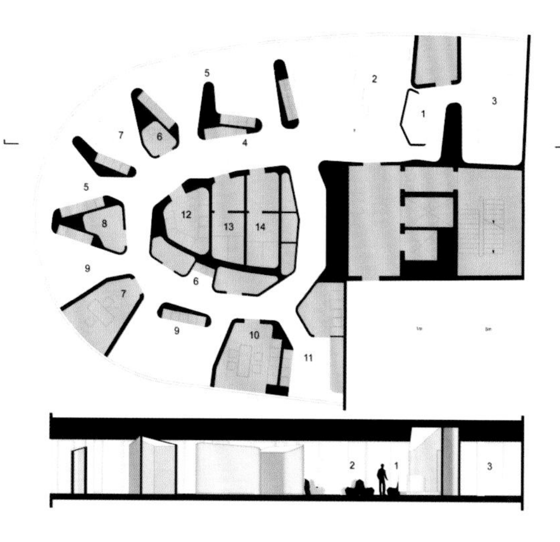

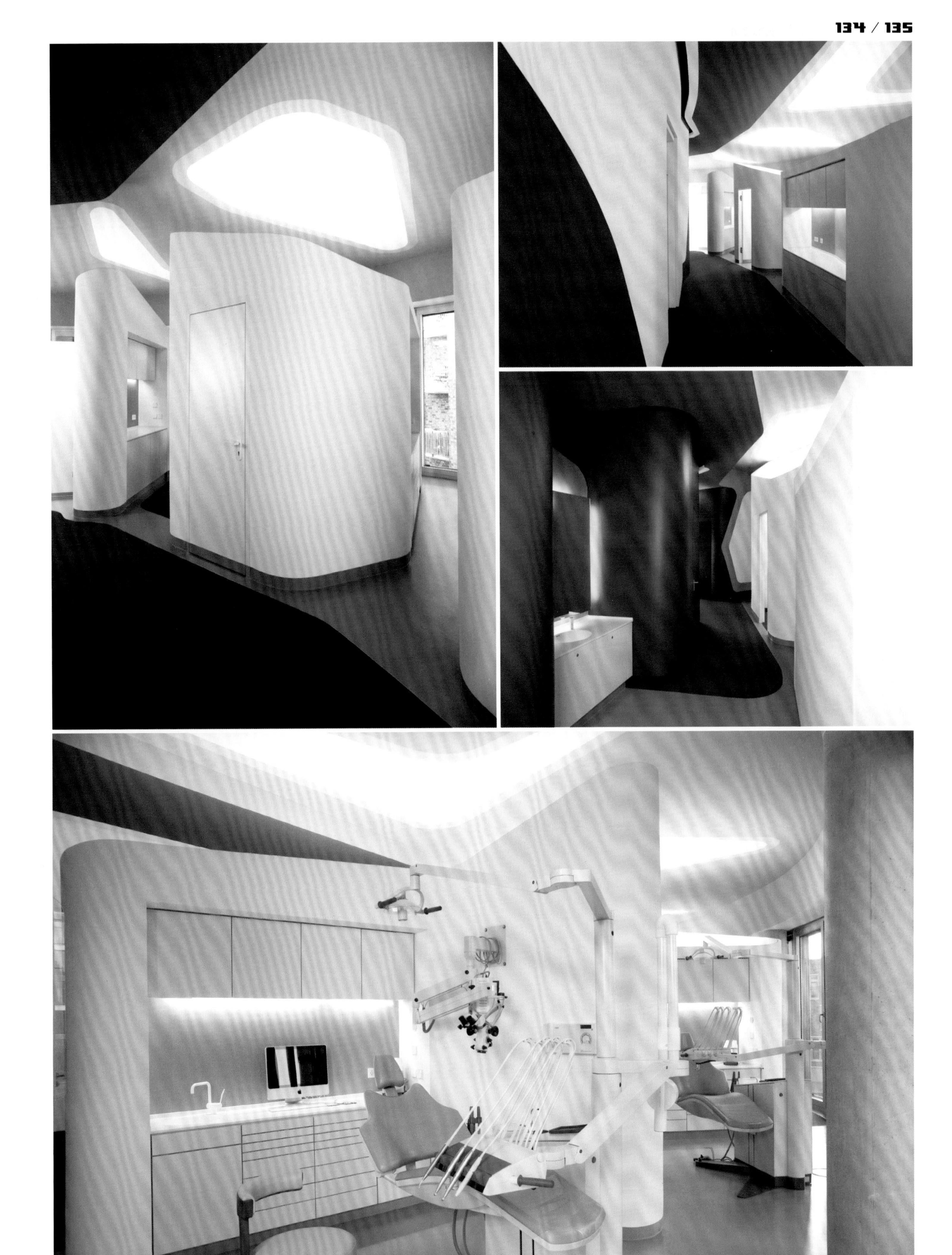

上海安娜贝尔SPA会所

SHANGHAI ANNABEL SPACLUB

项目资料：
设计单位：山村富弘室内设计工程有限公司
设计师：周家永
主要材料：柚木、碳烤木木地板、海贝石、沙雕、藤编
项目地址：上海沪青平公路
建筑面积：1132m^2

Project Information:
Design Unit: Shancunfuhong Interior Design Engineering Co., Ltd.
Designer: Zhou Jiayong
Materials: Teak, BBQ Wood Flooring, Seashells Stone, Sand Sculptures, Rattan
Project Address: Shanghai Luqingping Road
Area: 1132sqm

安娜貝爾
Spa 生活館

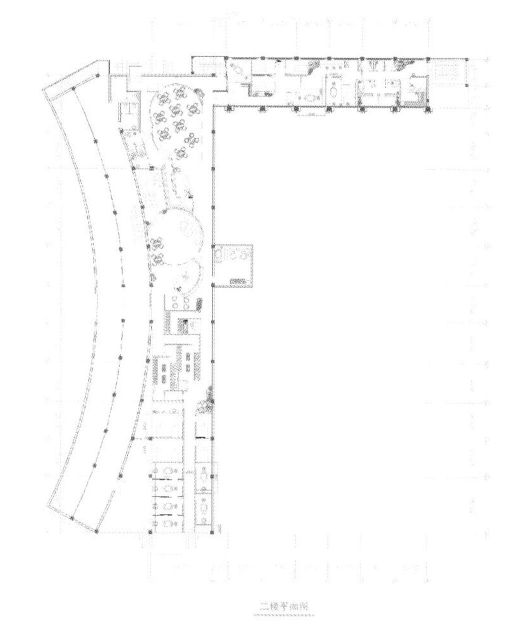

二楼平面图

P69
P68
P75
P73
P80
P79
P85
P86

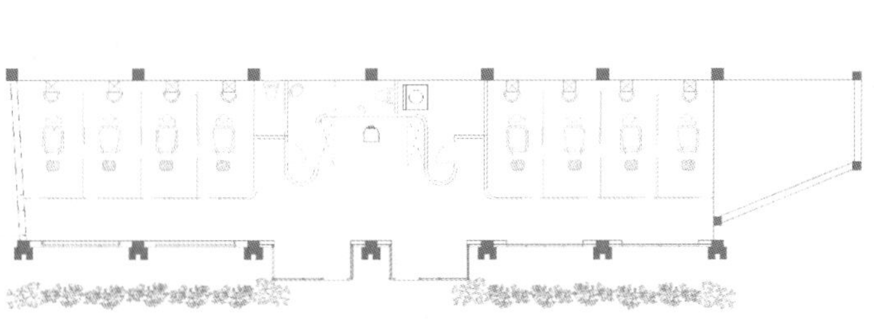

一楼平面图

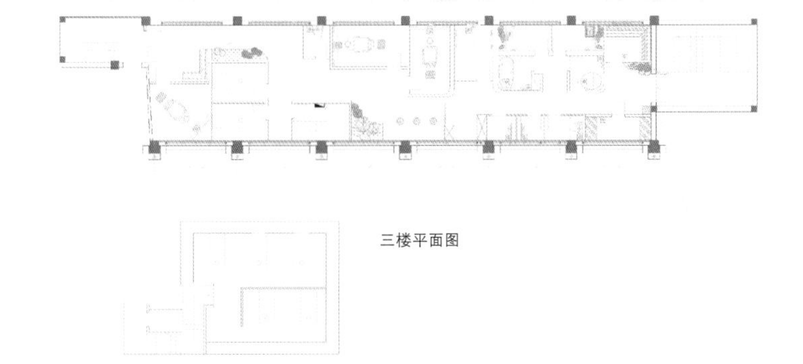

三楼平面图

MIRA SPA

MIRA SPA

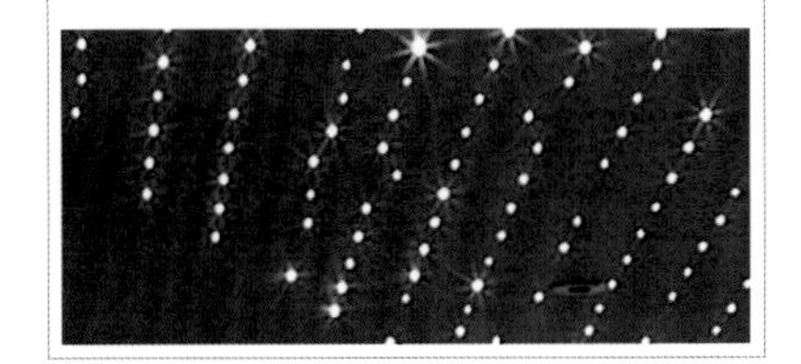

项目资料：
设计单位：CI3 Architects Limited.
客户：Mira Hotel
项目地址：香港
面积：1,612m²
完成时间：2009年

Project Information:
Design Unit: CI3 Architects Limited.
Client: Mira Hotel
Project Address: Hong Kong
Area: 1,612 sqm
Completion: 2009

侨治美容美发连锁杭州庆春路银泰店

GEOGE BEAUTY AND HAIRSTYLE CHAIN HANGZHOU QINGCHUN ROAD YINTAI SHOP

项目资料：

设计单位：杭州屹展室内设计工作室
设计总监（主创）：蒋捍明 肖懿展
参与设计团队：胡艳杰 黄道永
摄影师：訾向平
项目地址：杭州庆春路银泰
面积：180m²
主要材料：浅色高密度板
完成时间：2010年1月

Project Information：

Design Unit：Hangzhou Yizhan Interior Design Studio
Design Director：Jiang Hanming, Xiao Yizhan
Involved Design Team：Hu Yanjie, Huang Daoyong
Photographer：Zi Xiangping
Project Address：Hangzhou Qingchun Road Yintai
Area：180 sqm
Materials：Light-colored high density Broad
Completion：January, 2010

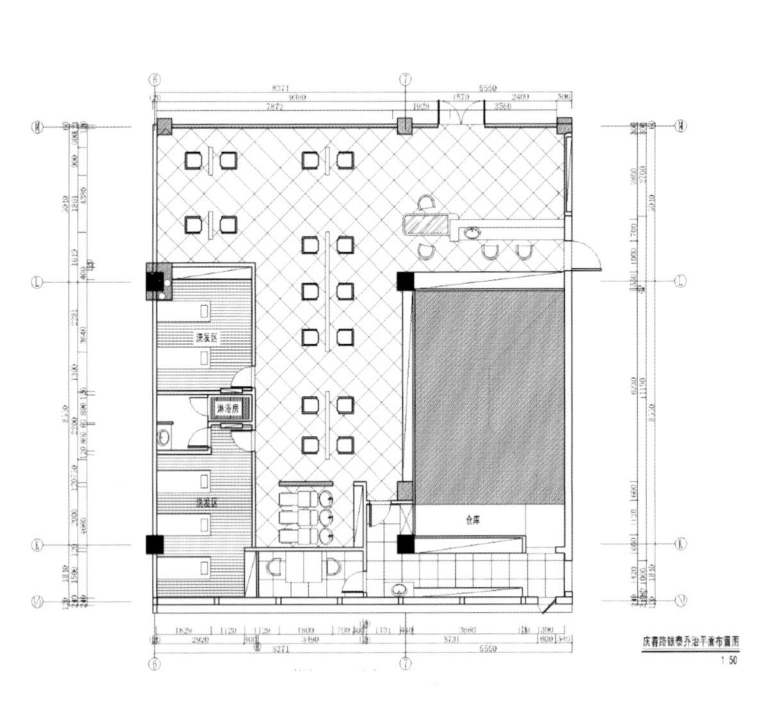

庆春路银泰乔治平面布置图
1:50

YI-SPA-STUDIO

YI-SPA-STUDIO

项目资料：
设计单位：Plajer & Franz Studio
项目管理：kathrinriet schel
摄影师：ken schluchtmann
客户：xix gesellschaft mbh
面积：120m²

Project Information:
Design Unit: Plajer & Franz Studio
Project manager: kathrinriet schel
Photographer: ken schluchtmann
Client: xix gesellschaft mbh
Area: 120sqm

CLUB
会所

MPC HIRES

MPC HI RES

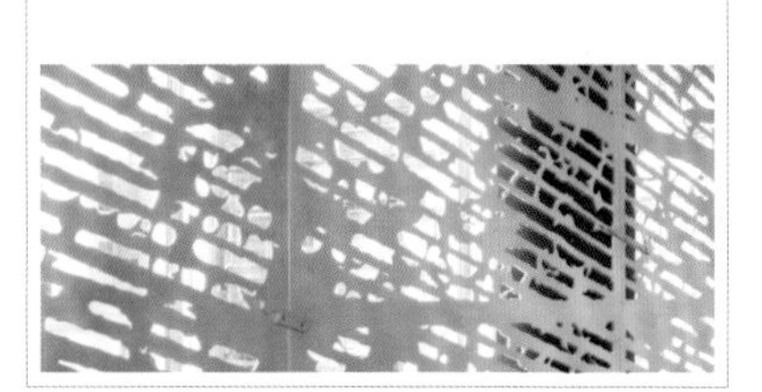

项目资料：
设计单位：Moving Picture Company

Project Information：
Design Unit：Moving Picture Company

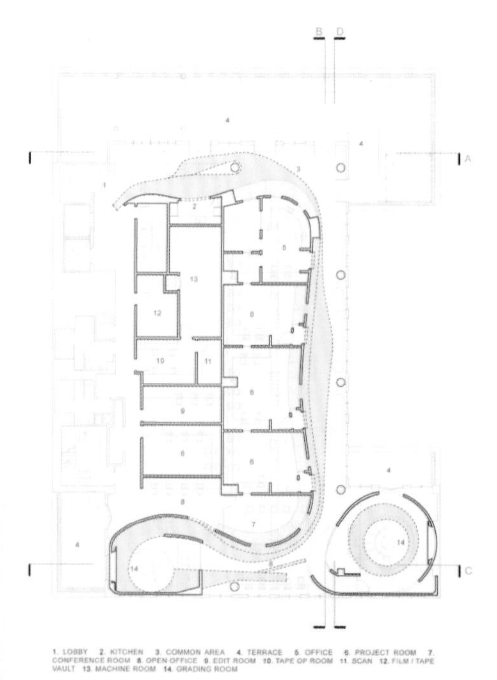

1. LOBBY 2. KITCHEN 3. COMMON AREA 4. TERRACE 5. OFFICE 6. PROJECT ROOM 7. CONFERENCE ROOM 8. OPEN OFFICE 9. EDIT ROOM 10. TAPE OP ROOM 11. SCAN 12. FILM / TAPE VAULT 13. MACHINE ROOM 14. GRADING ROOM

国美建设“华尔道夫II”公共设施

GUOMEI CONSTRUCTION “WALDARF II” PUBLIC FACILITES

项目资料：

设计单位：动象国际室内装修有限公司
设计师：谭精忠
参与设计团队：詹智惟
项目地址：台北市长春路
面积：929m²
主要材料：西雅图石材、棕灰石、雪花石、马毛皮、镀钛、橡木染灰、墨镜、茶玻、施华洛世奇水晶灯

Project Information:

Design Unit: Dongxiang International Interior Decoration Co., Ltd.
Designer: Tan Jingzhong
Involed Design Team: Zhan Zhiwei
Project Address: Taipei Changchun Road
Area: 929sqm
Materials: Seattle stone, brown limestone, alabaster, horse fur, titanium, oak dyed gray, dark glasses, tea glass, Swarovski Crystal Light

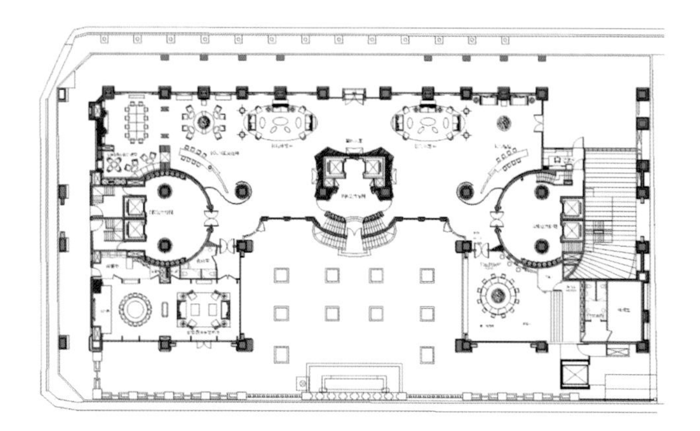

城南逸家天穹会所

NORTH OF TOWN YI HOUSE-TIANQIONG CLUB

项目资料：
设计单位：水平线空间设计
设计总监（主创）：琚宾
参与设计团队：张轩崇 石燕 尹芮
摄影师：孙翔宇
项目地址：成都
主要材料：铜色肌理漆、金世纪、蝴蝶绿、雅士白

Project Information:
Design Unit: Level Space Design
Design Director(Main Director): Ju Bin
Involed Design Team: Zhang Xuanchong, Shi Yan, Yin Rui
Photographer: Sun Xiangyu
Project Address: Chengdu
Materials: Bronze fabric paint, golden century, Butterfly Green, Aston White

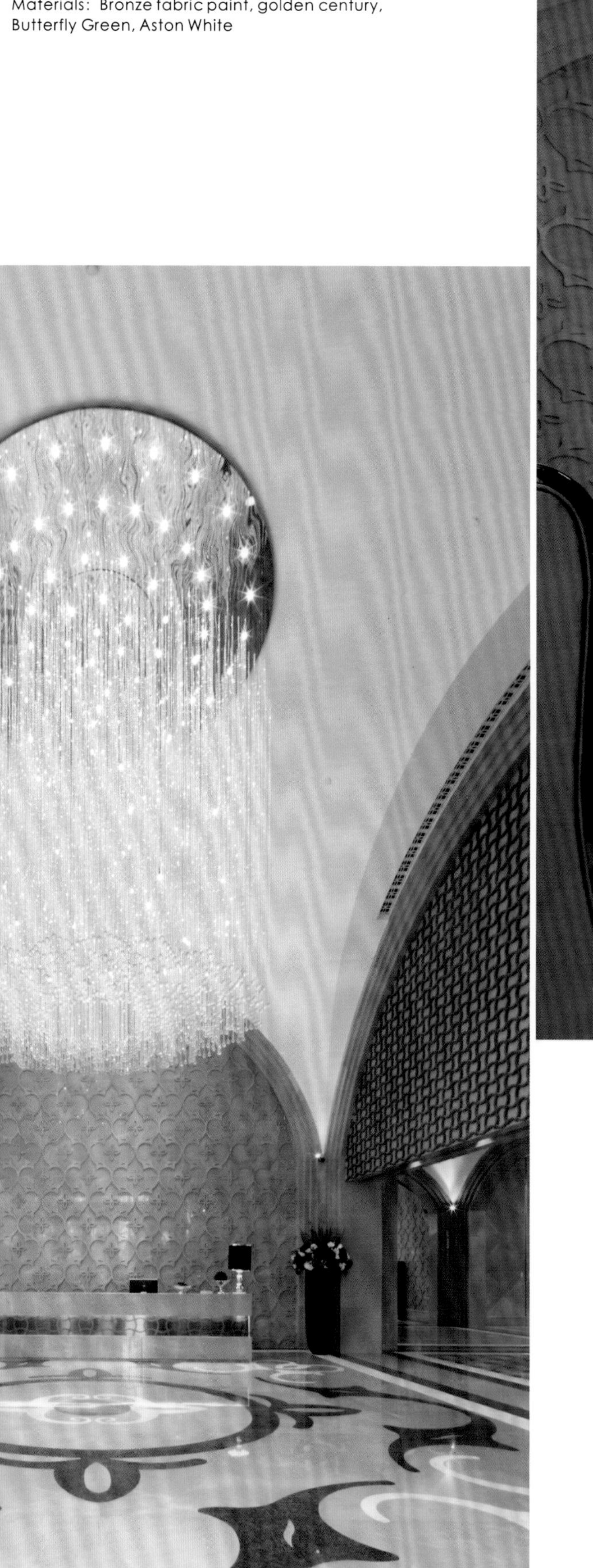

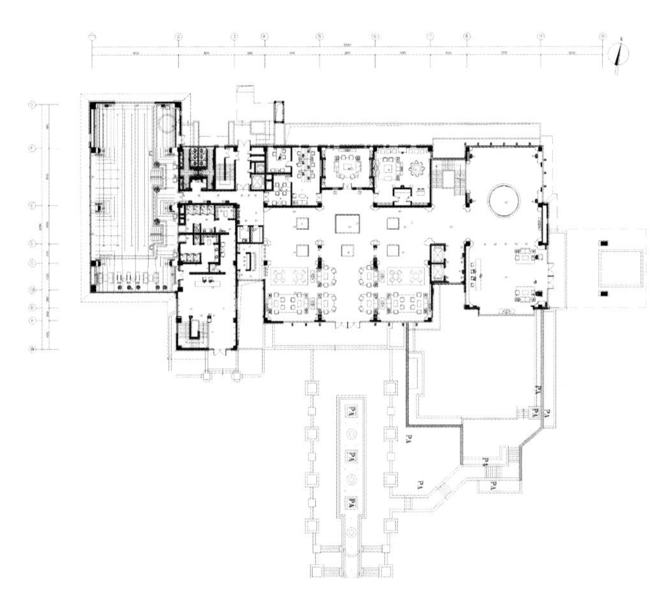

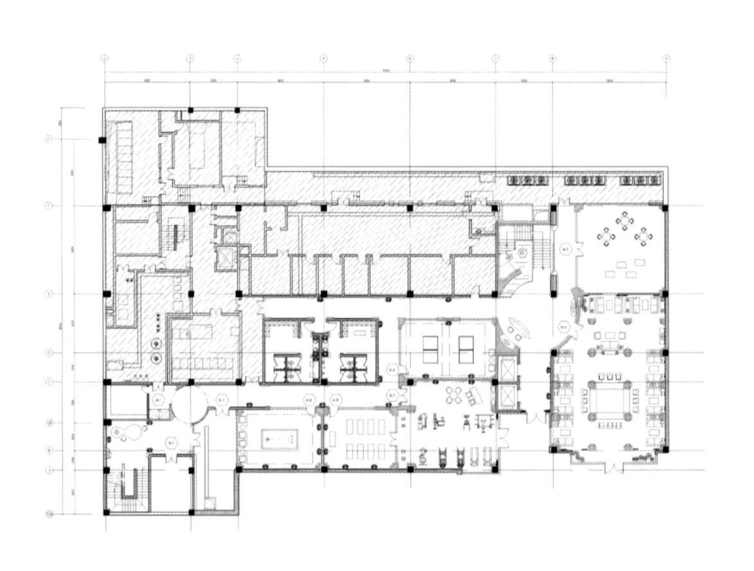

CITY CENTER'S ARIA POOL DECK

CITY CENTER'S ARIA POOL DECK

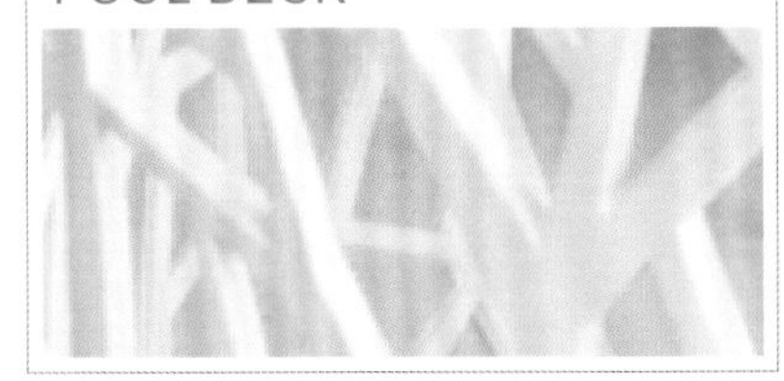

项目资料：
GRAFT设计团队：Lars Krückeberg, Wolfram Putz, Thomas Willemeit, Alejandra Lillo
Project Manager：Brian Wickersham
设计师：Lorena Yamamoto, Rob Decosmo, Kenneth Cameron,Elizabeth Pritchett, Brian D Nelson, Cornelia Faisst, Joshua Gilpin, Andrea Schutte
Leo Kocan, Seyavash Zohoori, Stefan Grohne, Casey Rehm, Nathan Miller, Asami Tachikawa, Naoko Miyano, Christoph Korner, Lillian Yang, Elizabeth Wendell, Ian Ream, Juyen Lee, Christoph Jantos, Brandon Love, Celi Freeman, Stefan Beese, Frank Lin
摄影师：Ricardo Ridecos
项目地址：USA, Nevada, Las Vegas - Aria Resort and Casino
面积：15793m²
完成时间：2010年3月

Project Information：
GRAFT Design Team：Lars Kruckeberg, Wolfram Putz, Thomas Willemeit, Alejandra Lillo
Project Manager：Brian Wickersham
Designers：Lorena Yamamoto, Rob Decosmo, Kenneth Cameron,Elizabeth Pritchett, Brian D Nelson, Cornelia Faisst, Joshua Gilpin, Andrea Schutte Leo Kocan, Seyavash Zohoori, Stefan Grohne, Casey Rehm, Nathan Miller, Asami Tachikawa, Naoko Miyano,Christoph Korner, Lillian Yang, Elizabeth Wendell, Ian Ream, Juyen Lee, Christoph Jantos,Brandon Love, Celi Freeman, Stefan Beese, Frank Lin,
Photographer：Ricardo Ridecos
Project Address：USA, Nevada, Las Vegas - Aria Resort and Casino
Area：15793 sqm
Completion：March 2010

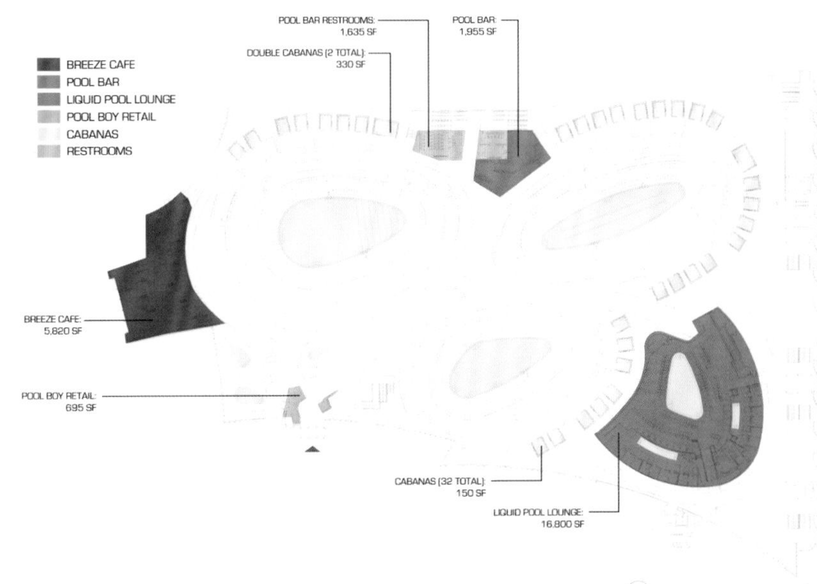
BREEZE CAFE
POOL BAR
LIQUID POOL LOUNGE
POOL BOY RETAIL
CABANAS
RESTROOMS
POOL BAR RESTROOMS: 1,635 SF
POOL BAR: 1,955 SF
DOUBLE CABANAS (2 TOTAL): 330 SF
BREEZE CAFE: 5,820 SF
POOL BOY RETAIL: 695 SF
CABANAS (32 TOTAL): 150 SF
LIQUID POOL LOUNGE: 16,800 SF
N

LIQUID

6

LIQUID
RESTROOMS

村前会馆

CLUB BEFORE THE VILLAGE

项目资料：
设计单位：上瑞元筑设计制作有限公司
设计师：冯嘉云、陈凤磊
摄影师：文宗博
项目地址：无锡市惠山区
面积：2700m²
主要材料：墙纸、灰麻、白影木、木格栅、大花白石材、老杨松风化白漆、黑钛
完成时间：2009年11月

Project Information:
Design Unit: Shangrui Yuanzhu Design Making Co., Ltd.
Designer: Feng Jiayun, Chen Fenglei
Photographer: Wen Zongbo
Project Address: Wuxi Huishan District
Area: 2700sqm
Materials: Wallpaper, gray linen, white wood, wood grille, large gray stone, old Yang pines of white paint, black titanium
Completion: November, 2009

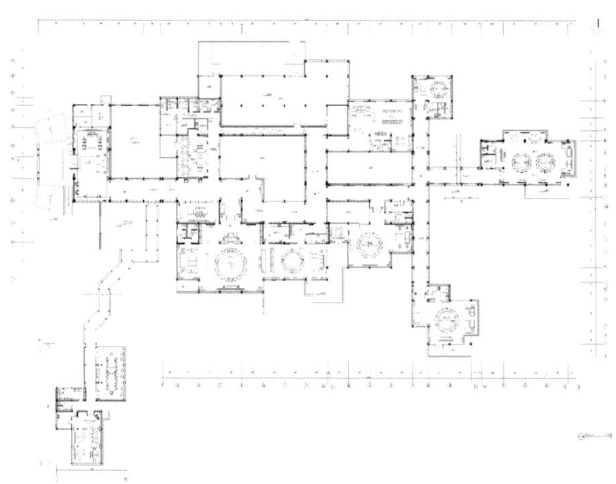

茗元会所

MINGYUAN CLUB

项目资料：

设计单位：福建国广一叶建筑装饰设计工程有限公司
设计师：何海华 李金珍
项目地点：福建 福州
面积：230m²
主要材料：玻璃、灰色软包、橡木擦黑色、镜面
完成时间：2009年9月

Project Information:

Design Unit: Fujian Guoguang Yiye Architectural Decoration Design Engineering Co., Ltd.
Designer: He Haihua, Li Jinzhen
Project Address: Fujian Fuzhou
Area: 230sqm
Materials: Glass, gray soft bag, black oak brush, mirror
Completion: September, 2009

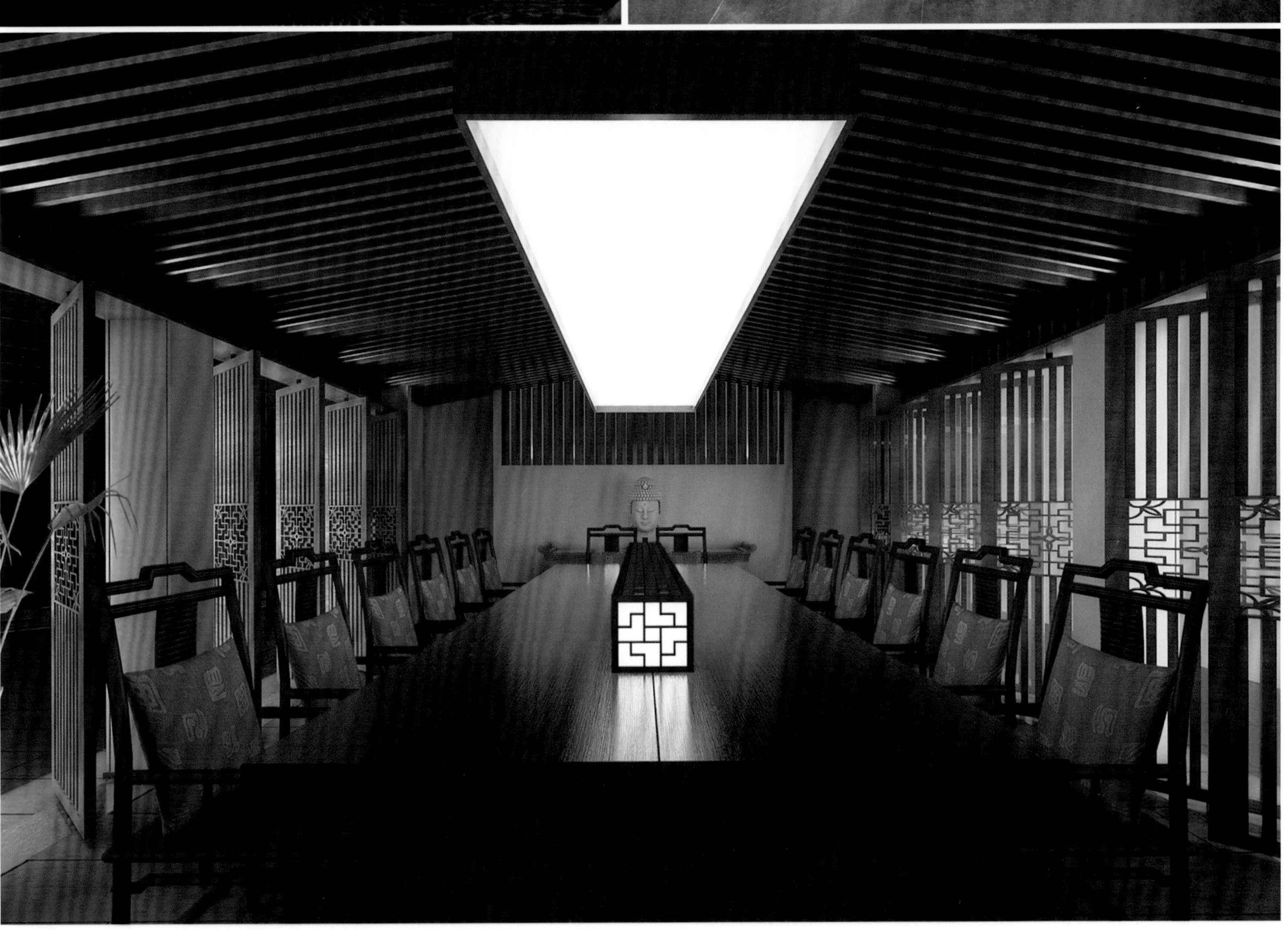

眼神2010会所

2010 EYES CLUB

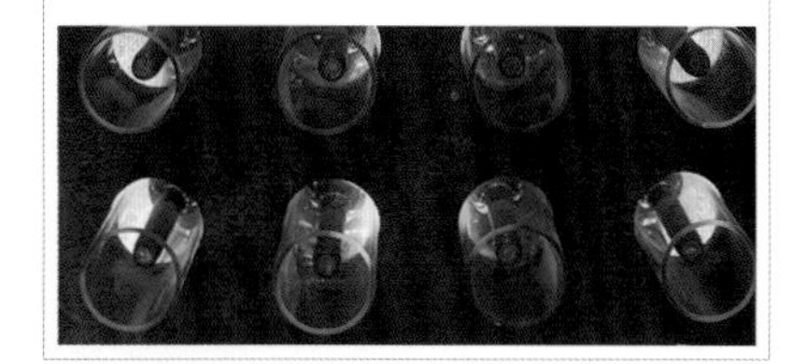

项目资料：

设计单位：佛山市大木明威社建筑工程设计有限公司
设计总监（主创）：谢智明
参与设计团队：何绮敏
摄影师：钱翔
项目地址：广东.佛山.创意产业园
主要材料：马赛克

Project Information:

Design Unit: Foshan Damumingwei Architectural Engineering Design Co., Ltd.
Design Director(Main Director): Xie Zhiming
Involed Design Team: He Qimin
Photographer: Qian Xiang
Project Address: Guangdong Foshan Creative Park
Materials: Mosaics

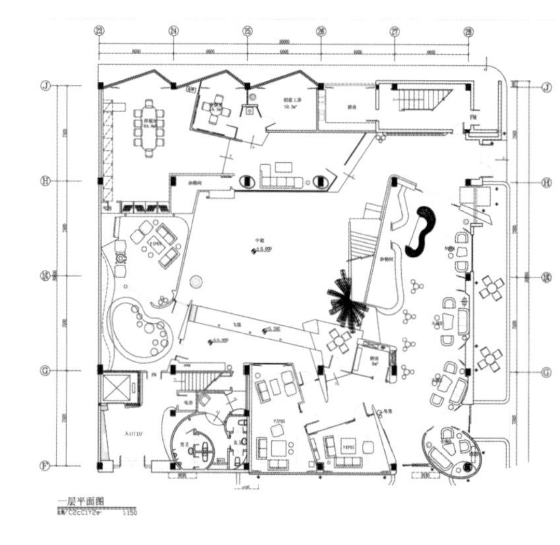

中国茶博汇

CHINA TEA MEETING

项目资料：

设计单位：福建国广一叶建筑装饰设计工程有限公司
设计师：何华武 陈亿元 何海华
项目地址：福建安溪
面积：2530m^2
主要材料：法国木纹石 白橡木擦色 红砖 银镜
山西黑花岗石 仿古砖 金属漆
完成时间：2010年2月

Project Information:

Design Unit: Fujian Guoguang Yiye Architectural Decoration Design Engineering Co., Ltd.
Designer: He Huawu, Chen Yiyuan, He Haihua
Project Address: Fujian Anxi
Area: 2530sqm
Materials: French wood Stone, white oak brush color, Red brick, silver mirror, Shanxi Black Granite Metallic tiles, antique brick, Metallic Paint
Completion: February, 2010

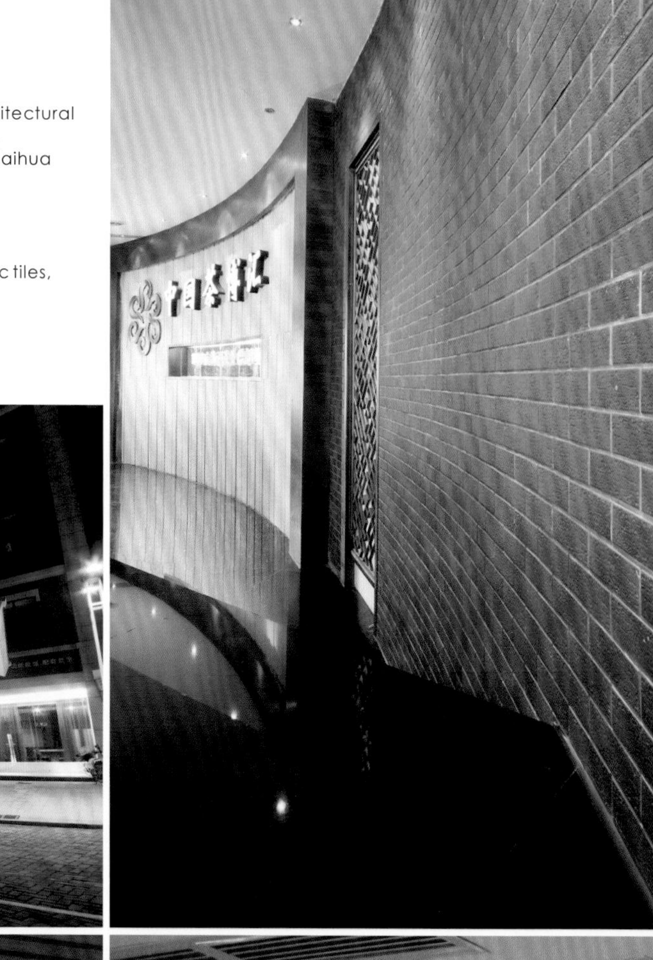

国茶博汇
1600亩海西茶业(总部)基地

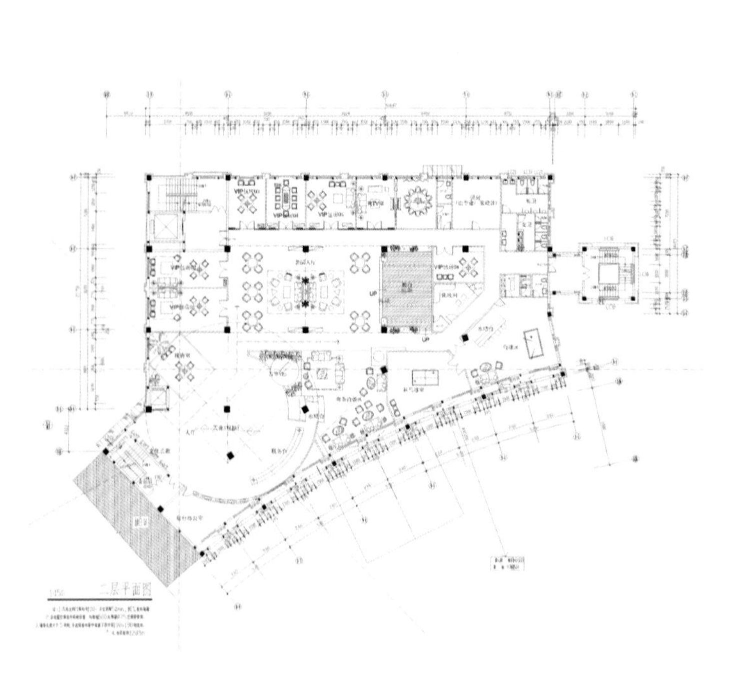

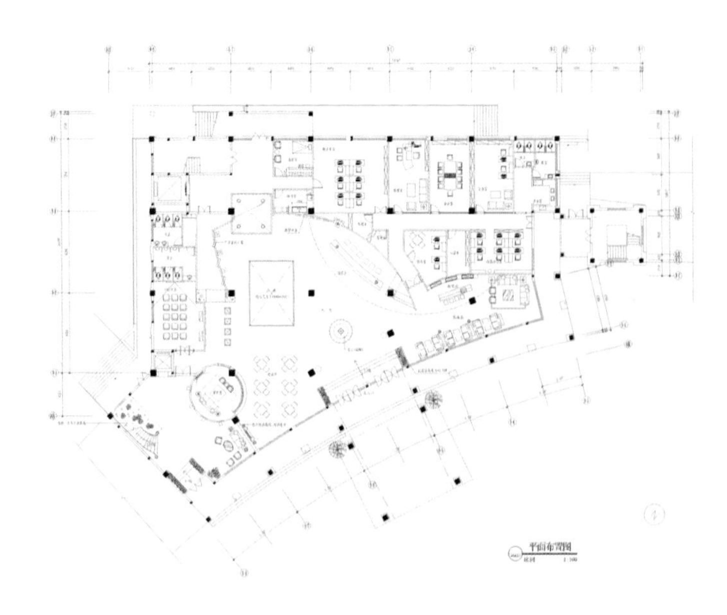

魅力四射的茶业特区
齐全配套 强大功能

SP075 THERME BAD RAGAZ

SP075 THERME BAD RAGAZ

项目资料：
设计单位：Smolenicky & Partner Architektur GmbH

Project Information:
Design Unit: Smolenicky & Partner Architektur GmbH

TAMINA
THERME
EINGANG

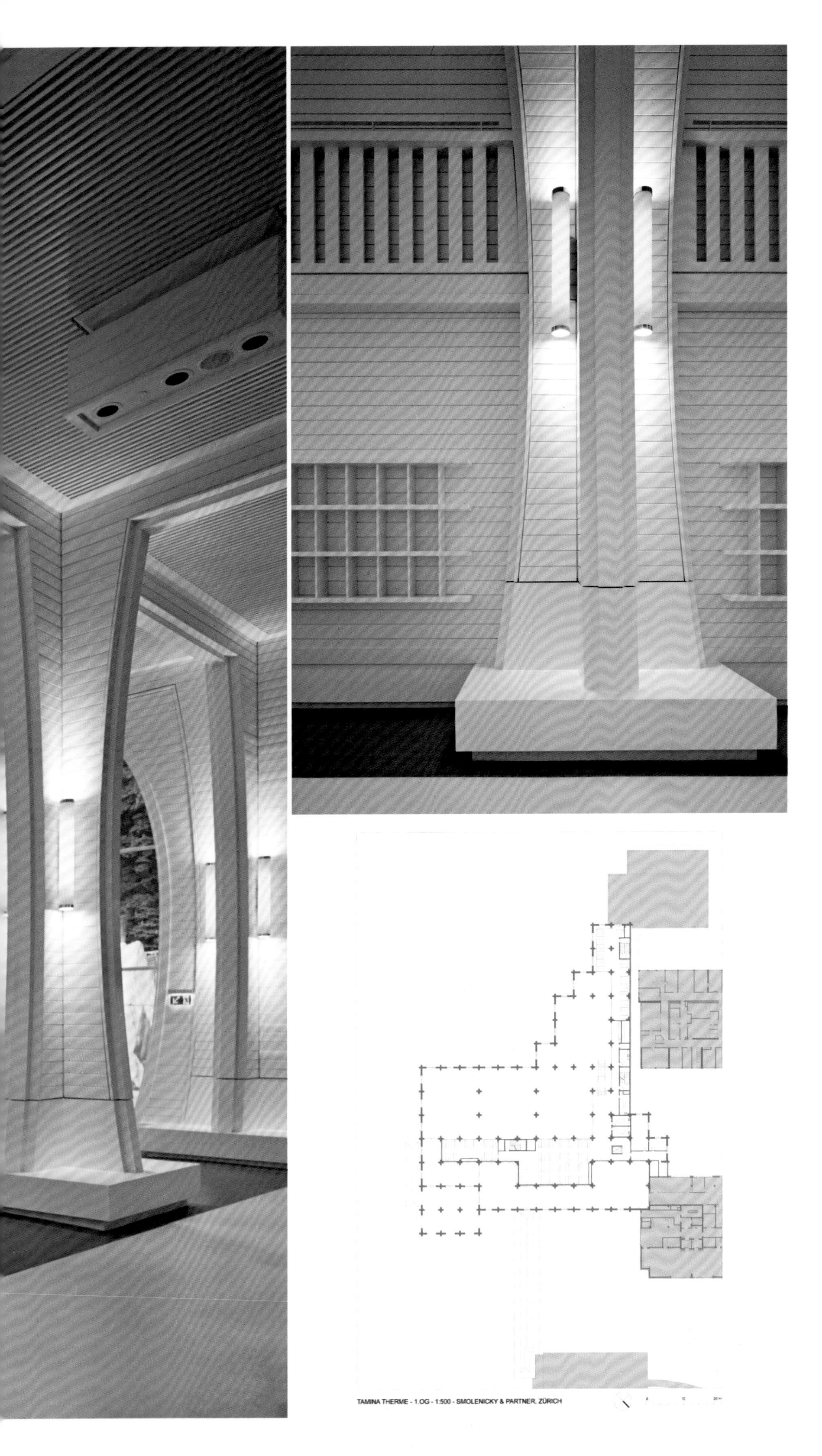

TAMINA THERME - 1.OG - 1:500 - SMOLENICKY & PARTNER, ZÜRICH

国美建设“隐哲”公共设施

GUOMEI CONSTRUCTION "YINZHE" PUBLIC FACILITES

项目资料：
设计单位：动象国际室内装修有限公司
设计师：谭精忠
参与设计团队：詹惠兰
项目地址：台北市金华街
面积：354m²
主要材料：铁刀木石材、黑云石、黑金锋、镀钛、铁刀木皮、灰玻

Project Information:
Design Unit: Dongxiang International Interior Decoration Co., Ltd.
Designer: Tan Jingzhong
Involed Design Team: Zhan Huilan
Project Address: Taipei Jinhua Street
Area: 354sqm
Materials: Chestnut trees stone, dark stone, black Jin Feng, titanium, iron knives, veneer, gray glass

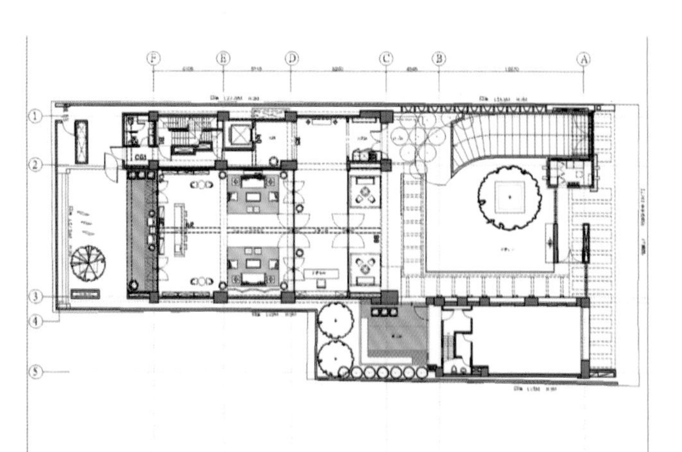

14 STREET Y RENOVATION

14 STREET Y RENOVATION

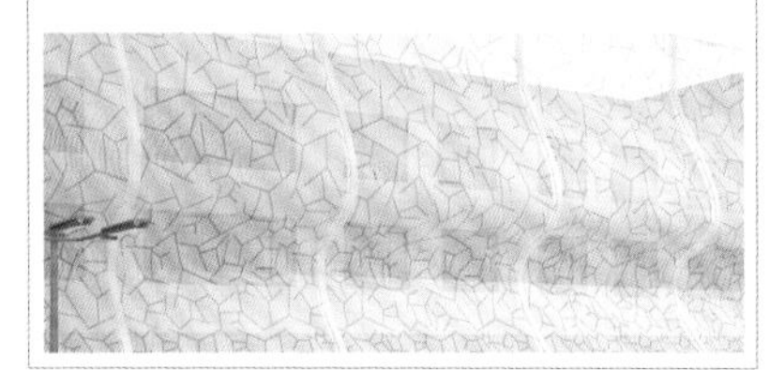

项目资料：

建筑与设计：Studio ST Architects & Z-A Studio
客户：The 14 Street Y of the Educational Alliance
项目管理：Séamus Henchy and Associates
MEP：Goldman Copeland Associates
声学工程师：Cerami and Associates
平面设计师：Rumors
承包商：Technetek LTD
面积：5556m²

Project Information:

Architecture and Design: Studio ST Architects & Z-A Studio
Client: The 14 Street Y of the Educational Alliance
Project Managers: Séamus Henchy and Associates
MEP: Goldman Copeland Associates
Acoustic Engineers: Cerami and Associates
Graphic Design: Rumors
Contractor: Technetek LTD
Area: 5556sqm

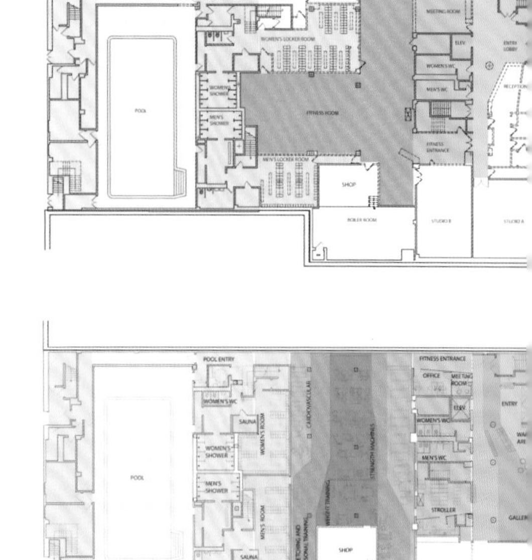
POOL ENTRY
WOMEN'S WC
SAUNA
WOMEN'S ROOM
WOMEN'S SHOWER
POOL
MEN'S SHOWER
MEN'S ROOM
SAUNA
MEN'S WC
CARDIOVASCULAR
STRENGTH MACHINES
WEIGHT TRAINING
STRETCHING AND PERSONAL TRAINING
SHOP
FITNESS ENTRANCE
OFFICE
MEETING ROOM
ELEV.
ENTRY
WOMEN'S WC
MEN'S WC
STROLLER
BOILER ROOM
STUDIO B
STUDIO A

AZLA

AZLA

项目资料：
设计单位：studio 0.10_[urban nodes]
摄影师：Benny Chan

Project Information:
Design Unit: studio 0.10_[urban nodes]
Photographer: Benny Chan

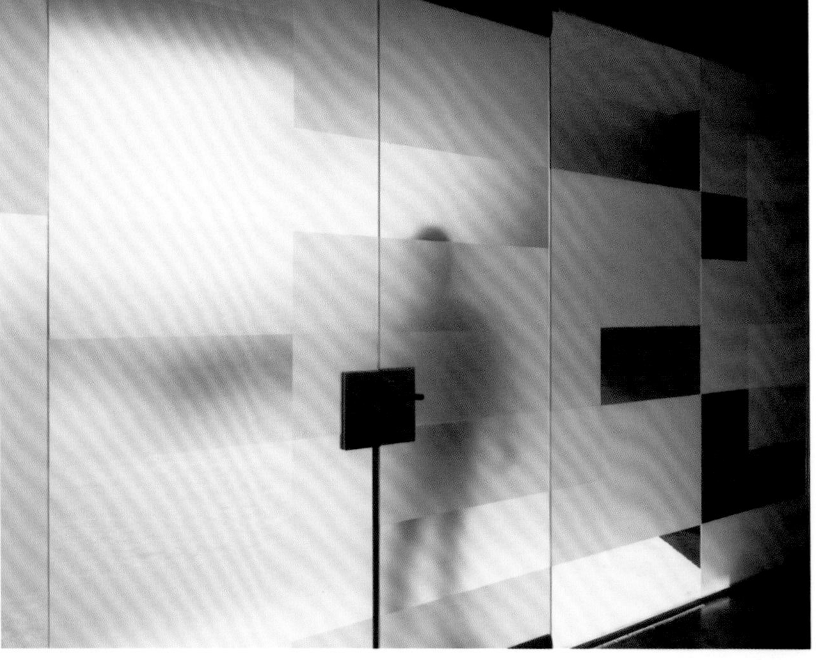

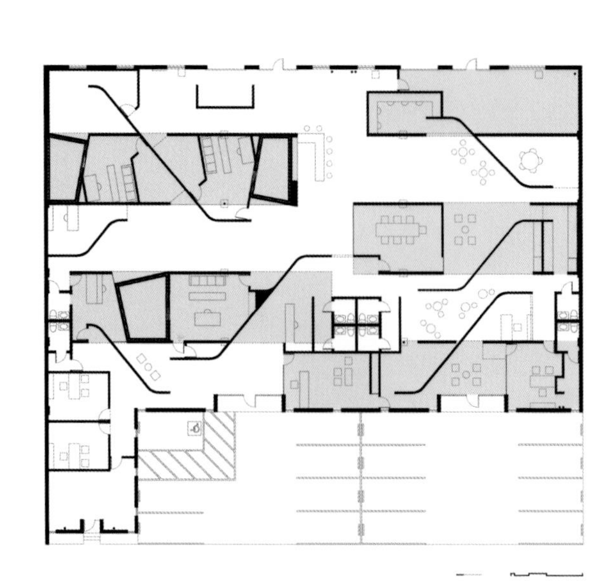

成都市都江堰玉垒锦绣高级商务会所

CHENGDU DUJIANGYAN YULEI JINXIU SENIOR BUSINESS CLUB

项目资料：

设计单位：四川大唐世家室内装饰设计有限公司
设计总监（主创）：黄河
参与设计团队：黄燚
摄影师：黄河 彭健
项目地址：成都市江堰离堆公园内

Project Information:

Design Unit: Sichuan Datangshijia Interior Decoration Design Co., Ltd.
Design Director(Main Director): Huang He
Involed Design Team: Huang Yi
Photographer: Huang He, Peng Jian
Project Address: inside Chengdu Jiangyan Lidui Garden

爱晚国汽车会所

AIWAN MOTOR CLUB

项目资料：

设计单位：杨凌志设计（香港）有限公司

Project Information:

Design Unit: YangLingZhi design (Hong Kong) limited

F&K

01
02
03
04
07
08
09
10

F&K

Aone

七匹狼武夷山会所

SEVEN WOLF WUYI MOUNTAIN CLUB

项目资料：

设计单位：福州林开新室内设计有限公司
设计师：林开新
摄影师：吴永长
项目地址：福建武夷山
面积：1000m²
主要材料：大理石、实木、木皮、玻璃、墙纸
完成时间：2008年

Project Information:

Design Unit: Fuzhou Linkaixin Interior Design Co., Ltd.
Designer: Lin Kaixin
Photographer: Wu Yongchang
Project Adress: Fujian Wuyi Mountain
Area: 1000 sqm
Materials: Marble, wood, veneer, glass, wallpaper
Completion: 2008

台中 BEING 健康俱乐部

BEING A HEALTH CLUB IN TAICHUNG

项目资料：

设计单位：鼎爵设计工程有限公司
设计总监（主创）：吕明颖
参与设计团队：黄千玳 张嘉璈 宋映燕 关嘉鸿
摄影师：李国民
项目地址：台中市忠明路
主要材料：清水模、钢构、玻璃、氧化镁板、南洋油樟木、柚木染色、麦饭石
面积：4955m²
开发商：统一佳佳股份有限公司

Project Information:

Design Unit: Dingjue Design Engineering Co., Ltd.
Design Director (Main Director): Lv Mingying
Involed Design Team: Huang Qiandai, Zhang Jia'ao, Song Yingyan, Guan Jiahong
Photographer: Li Guomin
Project Address: Taichung city loyal Ming road
Materials: Water mold, steel, glass, magnesium oxide board,
Nanyang oil camphor, teak stain, stone
Area: 4955sqm
Developer: Union Jiajia Co., Ltd.

BEING sport

BEINGsport
Vitality
www.beingsport.com
BEINGspa

BEING
BEING
Enjoy a
Relax
BEING
BEING

BEING
sport

台南BEING SPORT CLUB

TAINAN BEING SPORT CLUB

项目资料：
设计单位：鼎爵设计工程有限公司
设计总监（主创）：吕明颖
参与设计团队：张嘉璈 宋映燕 林义杰 钱欣 杨怡珍
摄影师：李国民
项目地址：台南市
主要材料：木作、钢构、石材、玻璃、水泥板、油樟实木地板、环氧树脂
项目面积：3304m²
开发商：统一佳佳股份有限公司

Project Information:
Design Unit: Dingjue Design Engineering Co., Ltd.
Design Director(Main Director): Lv Mingying
Involed Design Team: Zhang Jia'ao, Song Yingyan, Lin Yijie, Qian Xin, Yang Yizhen
Photographer: Li Guomin
Project Address: Tainan
Materials: Wood, steel, stone, glass, cement board, oil camphor wood flooring, epoxy resin
Area: 3304sqm
Developer: Union Jiajia Co., Ltd.

士邦健康俱

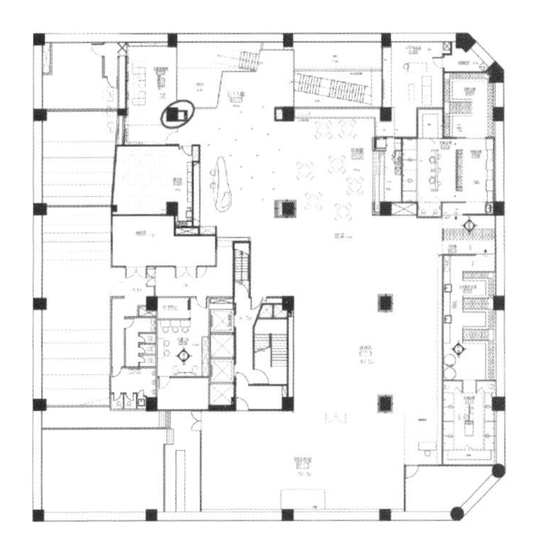

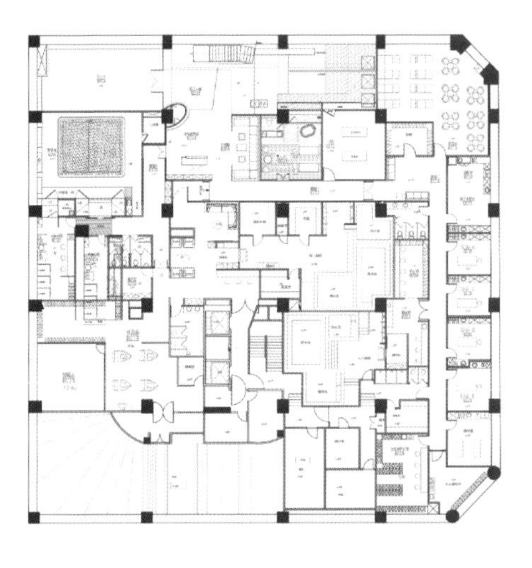

广州凯怡牙科会所

GUANGZHOU KAIYI DENTAL CLUB

项目资料：

设计单位：深圳任清泉设计有限公司
设计总监（主创）：任清泉
参与设计团队：王立春 马旭 柳俊伊
摄影师：任清泉
项目地址：广州市天河区北路2号冰花酒店11楼
主要材料：黑镜、仿古砖、木饰面、浅啡网、鸵油

Project Information:

Design Unit: Shenzhen Renqingquan Design Co., Ltd.
Design Director: Ren Qingquan
Involved Design Team: Wang Lichun, Ma Xu, Liu Junyin
Photographer: Ren Qingquan
Project Address: Floor 11 Ice Flower Hotel No. 2 North Road Tianhe District Guangzhou
Materials: Black mirror, antique brick, wood finishes, light brown network, ostrich oil

世欧上江城会所

SHI'OU SHANGJIANG CLUB

项目资料：

设计单位：深圳市品伊设计顾问有限公司
设计总监：刘卫军
参与设计团队：PINKI(品伊)创意机构
摄影师：文宗博
项目地址：中国福州
主要材料：银白龙大理石、夕化石、木饰面等

Project Information:

Design Unit: Shenzhen Pinyi design consulting Co., LTD
Design Director: Liu Weijun
Involed Design Team: PINKI Pinyi Creative Institution
Photographer: Wen Zongbo
Project Address: China Fuzhou
Materials: Silver-white Dragon marble, Xihua Stone, wood finishes

永旺岛会所

YONGWANG ISLAND CLUB

项目资料：
设计单位：江苏省吴江市清水装饰设计有限公司

Project Information:
Design Unit: Wujiang city of jiangsu province water adornment design Co., LTD

合一庭 · 中药医学联合会馆

ONE COURT · CHINESE MEDICAL ASSOCIATION

项目资料：
设计单位：香港东仓建设集团有限公司
设计总监（主创）：余霖
参与设计团队：李杰智 刘健 杨峰
项目地址：中国 成都 芙蓉长卷绿地产业项目一期
开发商：成都置信地产集团
面积：3000m²
主要材料：塑木、科技木、太白青石才等

Project Information:
Design Unit: Hong Kong Dongcang Construction Group Co., Ltd.
Design Director(Main Director): Yu Lin
Involed Design Team: Li Jiezhi, Liu Jian, Yang Feng
Project Address: China Chengdu Hibiscus Changjuan Greenland Real Estate Project Stage One
Developer: Chengdu Zhixin Real Estate Group
Area: 3000 sqm
Materials: Plastic-wood, Technology Wood, Taibai bluestone etc.

实拍照片
VIEW PICTURE

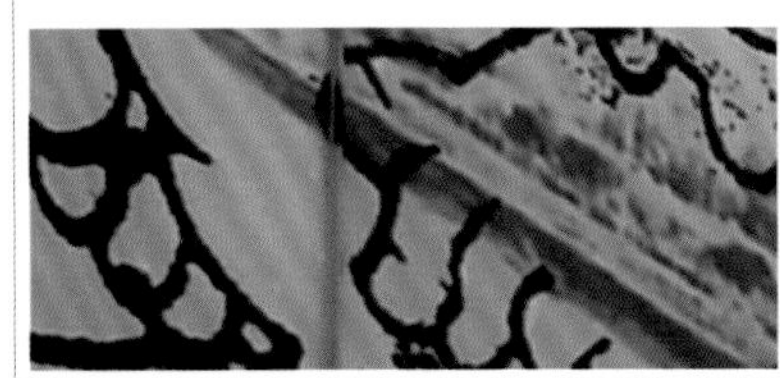

项目资料：
设计单位：广东卓艺设计顾问有限公司

Project Information:
Design Unit: Guangdong ZhuoYi design consulting Co., LTD

春
春

由申甲 清淞凯萨温泉会馆方案一

YOUSHENJIA QINGSONG KAISA HOT SPRING CLUB

项目资料：
设计单位：杨焕生建筑室内设计事务所
设计总监：杨焕生
参与设计团队：王慧静 郭士豪
摄影师：刘俊杰
项目地址：台北市

Project Information:
Design Unit: Yang Huansheng Architectural Interior Design Office
Design Director: Yang Huansheng
Involed Design Team: Wang Huijing, Guo Shihao
Photographer: Liu Junjie
Project Address: Taibei

由申甲 清淞凯萨温泉会馆方案二

YOUSHENJIA QINGSONG KAISA HOT SPRING CLUB

项目资料：
设计单位：杨焕生建筑室内设计事务所
设计总监：杨焕生
参与设计队团：王慧静 郭士豪
摄影师：刘俊杰
项目地址：台北市

Project Information：
Design Unit：Yang Huansheng Architectural Interior Design Office
Design Director：Yang Huansheng
Involed Design Team：Wang Huijing, Guo Shihao
Photographer：Liu Junjie
Project Address：Taibei

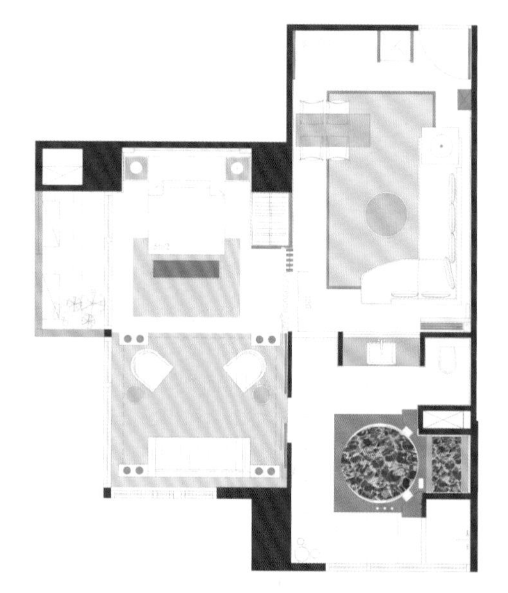

亚兰尼斯红酒会所

NICE WINE CLUBS SYRIA

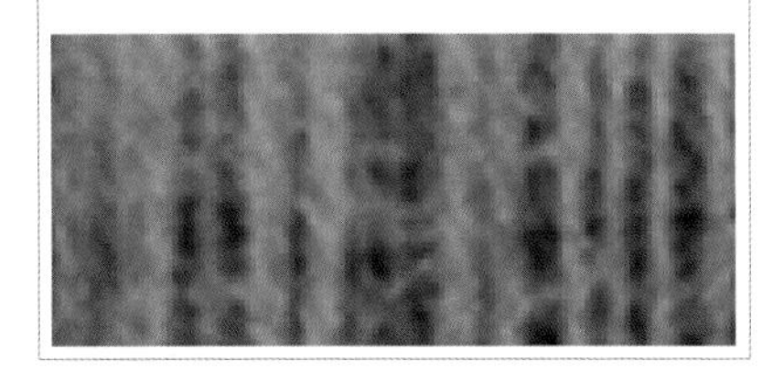

项目资料：

设计单位：浩澜设计事务所
设计总监（主创）：李浩澜
参与设计团队：朱国举
摄影师：李浩澜
主要材料：石材、灰镜、壁纸、乳胶漆、布艺等

Project Information:

Design Unit: Nanjing Haolan Design Office
Design Director(Main Director): Li Haolan
Involved Design Team: Zhu Guoju
Photographer: Li Haolan
Materials: stone, gray mirror, wallpaper, paint, cloth, etc

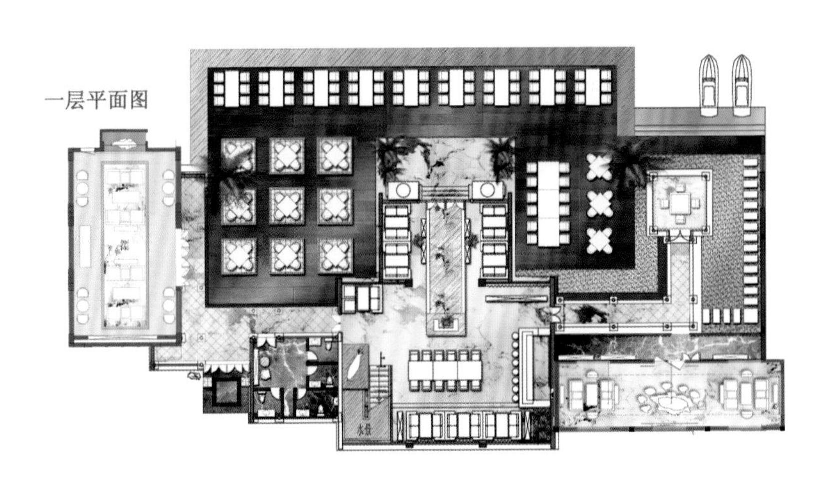
一层平面图

中联 · 江滨御景会所

ZHONGLIAN.JIANGBIN YUJING CLUB

项目资料：

设计单位：福建国广一叶建筑装饰设计工程有限公司
设计师：何海华 吴凤珍 高玲敏 吴华
项目地址：福建福清
面积：1800m²
主要材料：贵州木纹石、灰洞大理石、白色人造石、黑金沙大理石、福鼎黑火烧面、黑木纹石、15cm钢化磨砂玻璃、6cm大理石纹透光石、5cm茶镜玻璃、樱桃木面板
完成时间：2009年8月

Project Information:

Design Unit: Fujian Guoguang Yiye Architectural Decoration Design Engineering Co., Ltd.
Designer: He Haihua, Wu Fengzhen, Gao Lingmin, Wu Hua
Project Address: Fujian Fuqing
Area: 1800sqm
Materials: Guizhou Wood stone, Gray-cave marble, white artificial stone, black sands marble, Fuding black burned face, black wood stone, 15 cm steel and frosted glass, 6 cm marble translucent stone, 5 cm Tea mirror glass, cherry panel
Compeletion: August, 2009

01
PM-01
一层平面布置图
比例 1:100

中和圆通路许公馆

ZHONGHE YUANTONG ROAD
XU HOME

项目资料:

设计单位：鼎爵设计工程有限公司
设计总监（主创）：吕明颖
参与设计团队：林义杰 杨怡珍 林雯萱
摄影师：林国民
项目地址：台北县 中和市圆通路
主要材料：清水模墙、空心砖墙、观音石、铁刀木染灰、麦饭石、鑚泥板、H型钢构、玻璃
面积：1387m^2
开发商：许先生

Project Information:

Design Unit: Dingjue Design Engineering Co., Ltd.
Design Director(Main Director): Lv Mingying
InvoIed Design Team: Lin Yijie, Yang Yizhen,Lin Wenxuan
Photographer: Lin Guomin
Project Address: Taipei Zhonghe City Yuantong Road
Materials: Water mold walls, hollow walls, Guanyin stone, iron knives stained ash wood,
stone, clay tablets Diamond, H-type steel, glass
Area: 41387sqm
Developer: Mr. Xu

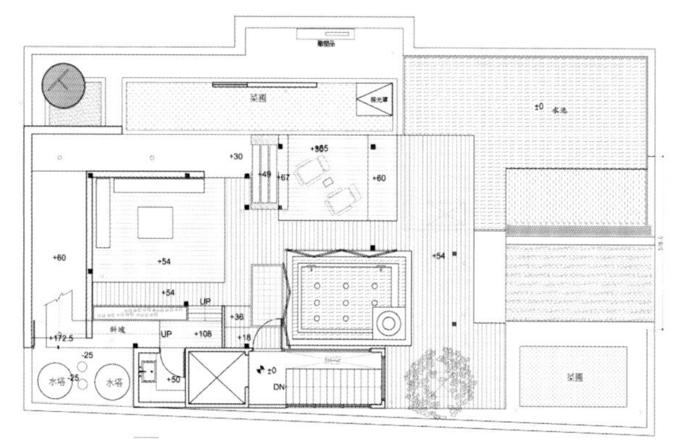

中和圆通路RF平面图 S: 1/80

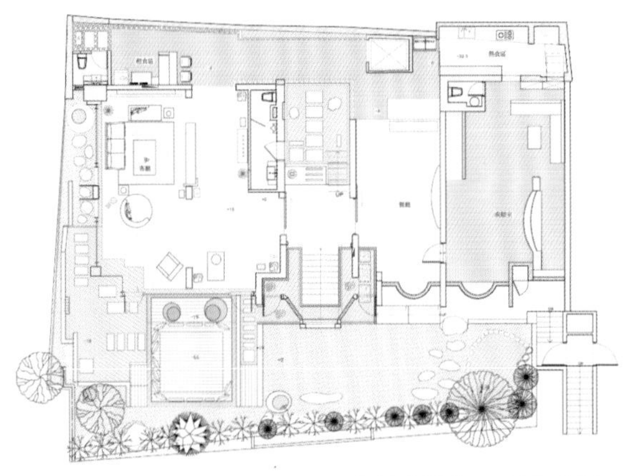

中和圆通路1F平面图 S: 1/80

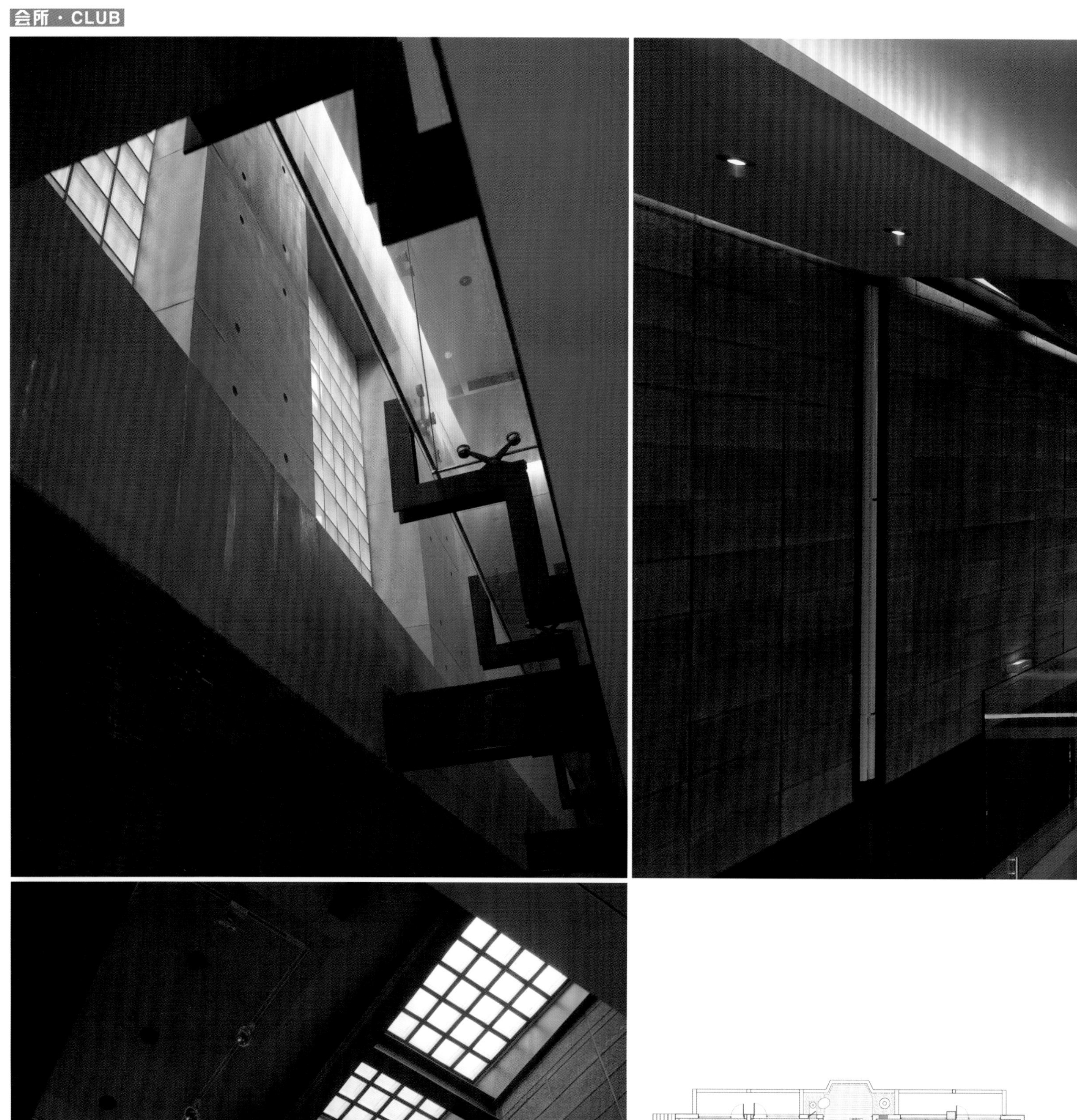

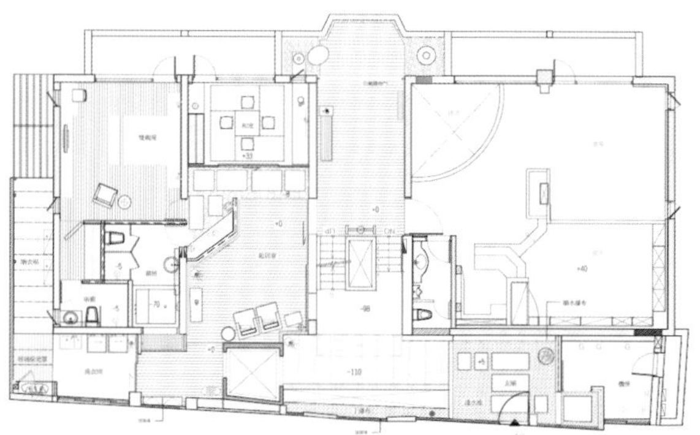

中和圆通路2F平面图 S: 1/80

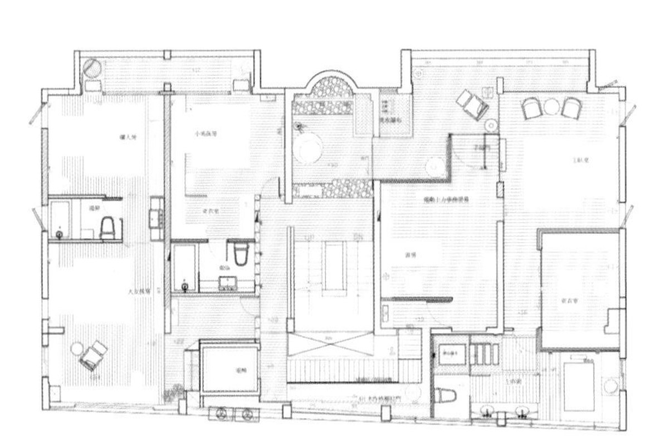

中和圆通路3F平面图 S：1/80

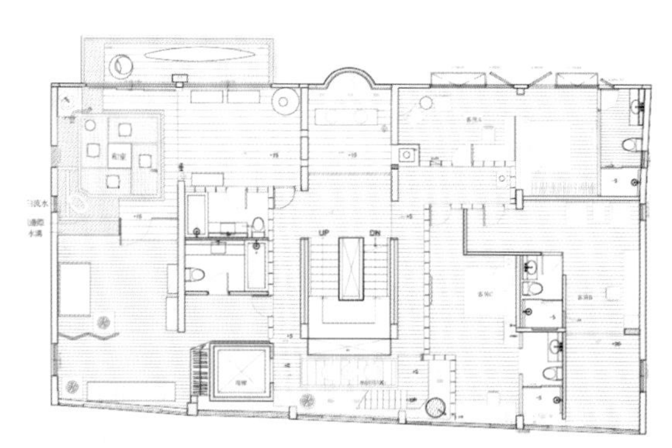

中和圆通路4F平面图 S: 1/80

常州香格里拉琉璃时光SPA会所

CHANGZHOU XIANGGELILA GLASS TIEMS SPACLUB

项目资料：

设计单位：深圳市黑龙室内设计有限公司
设计总监（主创）：王黑
参与设计团队：黑龙设计
摄影师：文宗博
主要材料：木纹石、中国黑、黑板岩、皇金凹凸板、瓷砖、斑马木、马赛克、胶地板、防火板、墙纸
面积：3200万m^2
项目地址：江苏常州市新北区通江大道398号
完成时间：2009年8月

Project Information:

Design Unit: Shenzhen Black Dragon Interior Design Co., Ltd.
Design Director: Wang Hei
Involved Design Team: Black Dragon Design
Photographer: Wen Zongbo
Materials: Wood Stone, China black, blackboard rock, imperial gold embossing plates, tiles, zebra wood, mosaic, rubber flooring, fire board, wallpaper
Area: 3200 Millionsqm
Project Address: Jiangsu Changzhou New North District Tongjiang Avenue No. 398
Completion: August, 2009

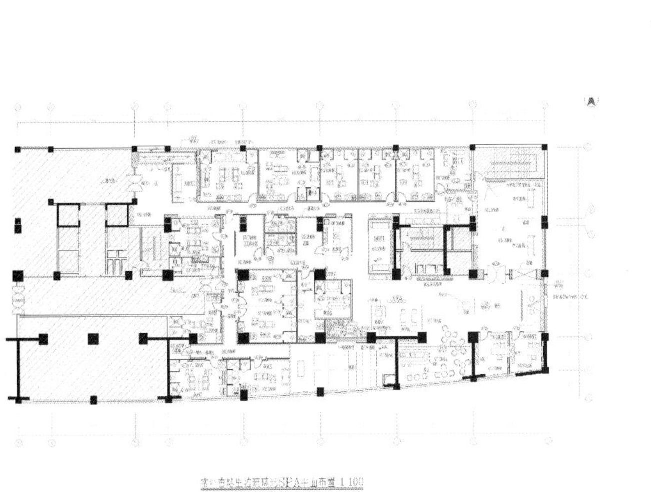
SPA平面布置 1 100